*Proceedings in Life Sciences*

21.3.87 - 20.3.88
**EUROPÄISCHES UMWELTJAHR**

Gisela Grupe and Bernd Herrmann (Eds.)

# Trace Elements in Environmental History

Proceedings of the Symposium held from
June 24th to 26th, 1987, at Göttingen

With 39 Figures

Springer-Verlag
Berlin Heidelberg New York
London Paris Tokyo

Dr. Gisela Grupe

Prof. Dr. Bernd Herrmann

Institut für Anthropologie der
Georg-August-Universität
Bürgerstraße 50
3400 Göttingen, FRG

Supported by Stiftung Volkswagenwerk

ISBN-13: 978-3-642-73299-7 e-ISBN-13: 978-3-642-73297-3
DOI: 10.1007/978-3-642-73297-3


Softcover reprint of the hardcover 1st edition 1988

**PREFACE**

This book contains the contributions to an European symposium on "Trace Elements in Environmental History", held from June 24th to 26th at Göttingen, FRG. The conference was organised by the Institute of Anthropology of the Georg August-University in Göttingen. At first glance, it might be surprising that the organizers are anthropologists. But this is a result of change of paradigm prehistoric anthropology is facing at the time. For decades, population development and population processes in the past have been looked at in terms of morphology, thus describing the diversity of human populations by the outer appearance of the skeletal findings and by the reconstruction of population structures. The new approach concentrates less on how people in the past looked like, but moreover on what they did and how they lived. Thus, research is based on ecosystem-theories, and it aims on the evaluation of ancient ecological features and past man/environment relationships. Research is encouraged since anthropologists are asked a lot of questions by historians and social scientists, who became more and more interested in the history of every day's life. Prehistoric anthropology today focuses also on manners, habits, ways of life and environmental constituents as they can be traced from skeletal remains, which represent an important historical source.

The ecosystemic approach is promising since the experiences of daily life certainly influence human behaviour, life style and mentality, thus directing reproduction and therefore population development. In Göttingen especially, we were probably influenced too by "long lasting effects" of ecologic research at the Georgia Augusta, starting with Bergmann and Griesebach in the early 19th century.

Within this approach, trace element analysis of ancient human remains is a very useful tool in reconstructing conditions of life and environmental features, thus giving a simple answer to the question why this conference was organised by us. However, it was one of our aims to bring together those European scholars who are experienced in the field of trace elements in environment and environmental history, for to improve the exchange of information in this dynamic area of research.

Since there is a common agreement that a lot of basic research still has to be done, the purpose of the meeting was to discuss basic aspects of trace element analysis of excavated human remains. In particular, the trace element content of excavated skeletons is supposed to reflect dietary habits of prehistoric people, including the weaning age of young infants. It is further supposed to reflect some pathological conditions, especially heavy metal intoxication. Research is encouraged because as long as the skeletons are well preserved, trace element contents similar to physiological values known from modern clinical investigations are found. This holds also true for known age- and sex-differences of certain trace element contents. It is therefore believed that trace element analysis of ancient human skeletons reveals several important aspects of past man/environment relations which might help us in understanding today's conditions.

The conference itself was organised around four interlocking headings: aspects of analytical chemistry and problems of analysing biological hard tissues; the factors influencing trace element contents in ancient material during the time of interment, including experimental modelling of long-term phenomena; recent clinical data with respect to their diagnostic value for ancient material; and, of course, studies in prehistoric and historic materials. In particular, questions were asked on the kind of information provided by the trace element content of ancient hair and teeth, besides the far more frequent analysis of bone; the special treatments ancient materials need prior to analysis were mentioned; and we took a closer look both on the discrimination factors for various elements in the food chain and on more detailed information concerning palaeopathological diagnosis.

The answers to these questions can only be found by an interdisciplinary approach. We were therefore very pleased that scientists of various disciplines accepted our invitation. Thus anthropologists, analytical and physico-chemists, geochemists, mineralogists, archaeologists and physicians were discussing the above mentioned topics.

The symposium has been made possible by financial support of the Stiftung Volkswagenwerk. The National Committee for Environmental Affairs awarded the Logo of the European Year of Environment. The symposium was also sponsored by CHF Müller Comp and Perkin-Elmer Comp. Our special thanks are due to Mr. Günter Dege of the Stiftung Volkswagenwerk and to Dr. Dieter Czeschlik of Springer Verlag and of course to our colleagues participating in the conference.

Göttingen, October 31st 1987

Gisela Grupe
Bernd Herrmann

TABLE OF CONTENTS

**INVITED PARTICIPANTS**

Dr. Luc Buchet
Centre de Recherches Archéologiques/CNRS
Sophia-Antipolis,
F-06565 Valbonne Cedex

Prof. Dr. Klaus Dörner
Universitäts-Kinderklinik
Schwanenweg 20
D-2300 Kiel

Dr. Paolo Francalacci
Istituto di Antropologia e Paleontologia Umana
Via S Maria 55
I-56100 Pisa

Dr. Gisela Grupe
Institut für Anthropologie
Bürgerstr. 50
D-3400 Göttingen

Dr. Martina Hedrich
Hahn-Meitner-Institut Berlin GmbH
Glienicker Str. 100
D-1000 Berlin 39

Prof. Dr. Bernd Herrmann
Institut für Anthropologie
Bürgerstr. 50
D-3400 Göttingen

Dr. Bruno Kaufmann
Naturhistorisches Museum
Abt. Anthropologie
Augustinergasse 2
CH-4001 Basel

Dr. Simon Mays
Dept. for Archaeology
University of Southampton
UK-Southampton SO9 5NH

Prof. Dr. Theya Molleson
British Museum (Natural History)
Dept. for Palaeontology
Cromwell Road
UK-London SW7 5BD

Prof. Dr. Heinrich Newesely
Institut für Mineralogie und
Kristallographie der TU
Ernst-Reuter-Platz 1
D-1000 Berlin 12

Dr. Oskar Oster
Universitätsklinik Mainz
Abt. Klinische Chemie
Langenbeckstr. 1
D-6500 Mainz

Dr. Lex Runia
Instituut voor Prae- en Protohistorie
University of Amsterdam
Singel 453
NL-1012 WP Amsterdam

Dr. Iain Thornton
Imperial College of Science and
Technology, Royal School of Mines
UK-London SW7 2BP

Dr. Vincent Vernois
Laboratoire d'Histologie
Faculté de Chirurgie Dentaire
Paris V, 1 rue M. Arnoux
F-92120 Montrouge

Dr. Tony Waldron
31 Maidstone Road
Bounds Green
UK-London N11 2TR

Dr. Terry Williams
British Museum (Natural History)
Dept. of Mineralogy
Cromwell Road
UK-London SW7 5BD

Dipl.-Biol. Ursula Wittwer-Backofen
Anthropologisches Institut der
Universität, Saarstr.
D-6500 Mainz

Prof. Dr. Heinrich Zumkley
Medizinische Poliklinik
Albert Schweitzer Str. 33
D-4400 Münster

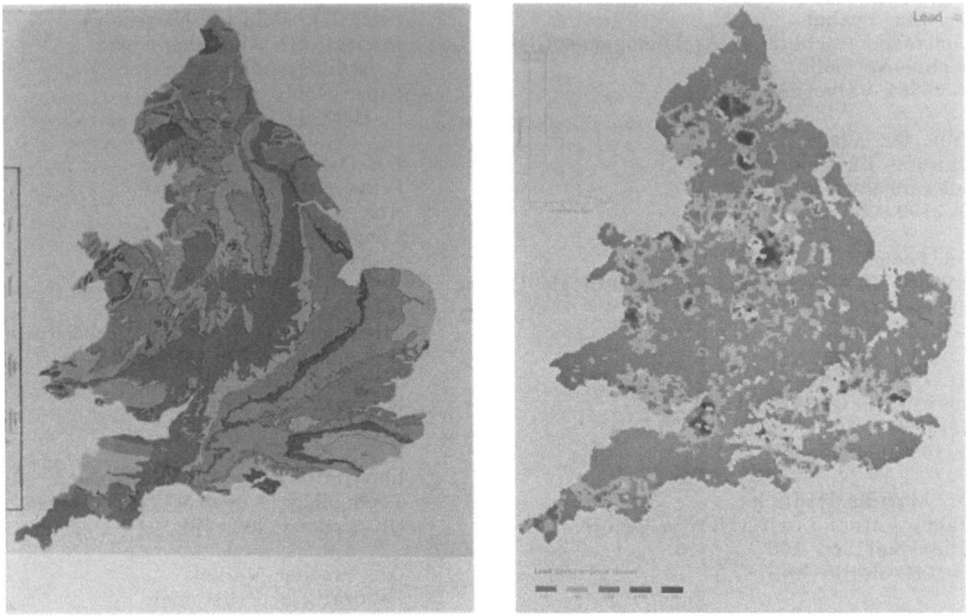

Plate 1: The distribution of lead in stream sediments in England and Wales (reproduced from Webb et al (1978) by permission of Oxford University Press. See contribution by I. Thornton, pp 135 - 144)

# CHEMICAL STABILITY OF HYDROXYAPATITE UNDER DIFFERENT CONDITIONS

Heinrich Newesely
Institut für Mineralogie und Kristallographie
Technische Universität Berlin
Ernst-Reuter-Platz 1
1000 Berlin 12, FRG

## Introduction

Recent investigations of the nature of chemical and biological processes with respect of environmental induced changes in biological hard tissues proceed beyond phenomenological and morphological observations to the question of whether and in what way the long term fate of bone and tooth mineral involve changes in the molecular and chemical structure and the stoichiometry of basic components. All previous observations demonstrate incontestably that many factors contribute to the initiation and spread of hard tissue decay and several to the transformation and recrystallization processes; nevertheless these processes have been clearly shown to be influenced specifically by certain chemical substances - the trace elements in particular.
Many of them have been discussed in this context (Li, Ag, Cu, Au, Hg, Sr, Cd, Zn, Be, Ba, Mg, B, Al, Y, Sn, Pb, Si, Se, Mo, Mn, Br, Fe, Co, Ni, Pt, Pd, F, and also pyrophosphate, carbonate, etc.), but we shall limit this review to a consideration of reaction partners and reaction conditions that influence the crystallization of apatite. The most important structural types in exchange reactions in the calcium phosphate system will be summarised.
The methods of investigation employed in this work to be described were - apart from the use of isotopes and physical measurements - study of crystal growth, chemical and microanalyses, spectroscopy, crystalchemical structural research, and electron microscopical morphological research. We shall consider first the basis of crystallization and nucleation, comparing the results obtained from geochemical investigations of apatite crystal formation in minerals with those from studies of formation and breakdown of calcium phosphates in biological systems. In a crystalchemical sense the effects of trace elements are mediated through structurally specific properties.

## Mineralogical/geochemical genesis

Apatite is very widespread in the mineral kingdom, although it participates only as an accessory in the make up of the earth's crust. Apatite crystals are found in magmatic rocks, in pegmatites, and in hydrothermally formed rocks. In sedimentary phosphorites poorly crystallized varieties of apatite are found, frequently with carbonate apatite as secondary formations. The growth of crystals is greatly influenced by geochemical processes near crystallization zones. For example, the well-known apatite ridge of Cerro de Mercado in Durange, Mexico, is demarcated by neighbouring iron ores; these apatite crystals are yellow to yellow-green and frequently contain inclusions such as magnetite and geothite. More important than these inclusions in the crystal chemical description of the crystals are substitutions in the domain of the lattice positions: for Ca→Sr, Pb, Na, Ce,..., for P→V, As, Si, S..., for OH→F, Cl,... The main problem of the crystal chemistry of the sparingly soluble calcium phosphates lies not so much in their systematics, but rather, because of the irregular structure even of single crystals (zone growth), in the description of pure single crystals, in concepts about the nature of poorly crystallized phases, and in predictions of stability toward hydrolysis and other transformations, as well.

Numerous varieties of apatite-like minerals have been classified, by measurements of elementary cells, into a comparatively simple system of substituted apatite forms. Strunz (1941), and Frondel (1943) pointed out that the greater part of some 17 mineral names appearing in the literature that describes calcium phosphates occuring in phosphorite deposits are actually only apatites, as well as whitlockite and hydrogen phosphates.

Hydroxyapatite crystallizes in the hexagonal dipyramidal crystal class 6/m. The pure crystals are colorless, have prismlike habitus, and are mostly isometric or needleformed.

The structure of hydroxyapatite shows the basis form of a structural series conforming to the apatite scheme $A_5X(YO_4)_3$. Hydroxyapatite must be fundamental to the structural series, since it is the simplest apatite, especially on crystallization from aqueous solutions, in that its components are limited to the system $CaO$-$P_2O_5$-$H_2O$. The possible limits of structural variations through substitution reactions can be clearly demonstrated in the apatite class: on the one hand, the phosphate lattice consists of a simple coordination lattice uncomplicated by cross-over elements, as in the silicates; on the other hand, a capacity exists for further variations through coordination of cations with simple anions or ion complexes. Rules for substitutions are based primarily on space considerations, symmetry factors, and polarization behaviour, as well as on the charge distribution in the lattice space of hydroxyapatite. The substituents replacing lattice components generally have the same valency as the exchanged ions. In coupled substitution, when valency differences of substituting cations or anions are compensated, e.g., in simultaneous exchange within the dimensional tolerances, as long

as electrical neutrality is preserved overall; for example: $2Ca^{2+}$ can be exchanged for $Na^{+}$ and $Y^{3+}$, or $2PO_4{}^{3-}$ for $SiO_4{}^{4-}$ and $SO_4{}^{2-}$.
Ions can replace each other isomorphically in the crystal lattice if the difference in ionic ratio does not exceed 15%: for example there exists an unlimited series of mixtures of fluorine and hydroxyapatites, or calcium and strontium apatite.
In the coupled exchange of calcium and phosphate in the apatite lattice, even ions whose sizes are beyond the limits already mentioned may achieve a stable lattice arrangement. The radius proportion $r_K/r_A$ remains constant within the stated limits (15%). Under these conditions the cation and anion partial lattices are greatly stretched;the total lattice, however, is in a tension-free state. Smaller ions cannot be introduced, even in coupled situations; they are subject to geometrical limitations by the spatial requirement of oxygen in the phosphate groups. An apatite with magnesium cannot therefore exist (ionic ratios of $Ca^{2+}$ = 0.99 Å; of $Mg^{2+}$ = 0.66 Å). Magnesium reacts with the calcium phosphate system in a different way: in precipitation reactions the whitlockite structure is induced in the presence of $Mg^{2+}$ ions. In the entry of substantially larger ions into the apatite lattice there is a definite apatite scheme for the cationic partial lattice as well as for the partial lattice of complexed anions, but the positions of univalent negative anions - assuming compensation of their charges in the total lattice by entry of low valent cations or high valent anions - may be partially or totally vacant.
For calcium apatites, anion-free sites of this kind are unknown: with shorter bond lengths in the calcium apatite lattice there is a corresponding increase in bond strength; also in the phosphate lattice, as has been mentioned, there is not a double possibility (above and below) of relative size distribution through coupled substitution. Cation-free sites may be discussed, but in the case of apatite they cannot be reconciled for compensation at astoichiometric composition.
The crystalchemical and micromorphological evaluations of preserved bone discoveries implicate therefore reliable methods as X-ray diffraction, electron microscopy, and spectroscopy with the skeletal materials and the soil environment in which bones are found. As specialisations of osteoarchaeometry become increasingly developed, so the need of new analytical techniques and tests of a skillful applicability becomes more necessary.
The reactivity of soils varies widely as geological and sedimentological conditions offer typical but different environments: gravels, chalk soil, clay, salt soils, sands, cave earths are examples of this wide variety, including atmospheric and biogenetic implications. The last mentioned features are strongly effective also in aride regions, with the well known fluctuations of high parching and dewiness. However, despite the diversity in depositions and burial modes only a few parameters govern the gradual decomposition of bone material:

1. pH-value of the surrounding medium;
2. humidity of the surrounding medium, may be governed directly by autolysis;

3. transport of matter, related to grain size, pore volume, solubility behaviour;
4. physical pressure;
5. destruction by microorganisms.

The reactivity of the bone fragments is shown in the afore named terms of exchange reactions within the crystal structure of the bone mineral, apatite $Ca_5(PO_4)_3OH$. This decomposition by substitution will often prevent the external appearance of buried bones. The extent of actual ion exchange with surrounding soil strata resp. transformation areas (sinter sheets) is to be shown quantitatively by X-ray fluorescence analysis; frequently a nearly total rearrangement of the anion lattice (phosphate, silicate vs. carbonate) as well as verified by scanning-electron microscopy and phase analysis by X-ray diffraction: calcite and quartz are the principal components of the sinters, additional diffuse apatite lines appear in bone samples. One may conclude the sinter formation to be fed as well from the bone part, as even the main bulk of sinter originate from the soil and the atmosphere. The bone substance disintegrates and physiological structures are dissolved by ion exchange, but by the ion exchange the outlines of skeletal fragments remain often preserved.

The wide variety of substitution in the mineralogical-chemical crystallization of apatite therefore to some extent is true for the biocrystallography of apatite; apatite formation under physiological conditions influenced particularly in the anion lattice by small cations.

Reactions of this nature will be discussed individually in the following sections.

## Biogenetic apatite formation

Although the elucidation of factors influencing apatite crystallization is most profitably studied using single crystals, in this context of a long lasting organisation and disintegration of textural units one must discuss the biological and/or sedimentological reaction conditions at, for example, precipitation crystallization.

Precipitation from neutral or alkaline aqueous solutions of calcium and phosphate ions have the composition and structure of apatite only under certain conditions; in general they form very fine, almost amorphous deposits. The crystallinity of the precipitates produced is influenced by the concentration of the reaction solutions: with large amounts of reactants, which yield huge quantities of precipitate, the amorphous form is almost stable and possesses properties differing from the crystalline state.

Because the precipitate is sparingly soluble, nucleation proceeds so fast that all the reactants are consumed in the process of forming nuclei and clusters. Whereas during crystallization reactions the nuclei are subject to dynamic equilibrium involving their coalescence and dissolution into simple ions, in precipitation reactions, because of their sparing solubility, the small phase nuclei are stable as such. Under the electron microscope, the smallest components of the precipitate appear as phase boundaries lacking definite outlines, often as poorly defined clusters measuring 50-100 Å in diameter. Under the conditions studied, the particle size of the calcium phosphate precipitate did not significantly depend on the pH of the solution. If the concentration of reactants is kept at lower levels, that is, if it approaches the solubility of the equilibrium conditions, then recrystallization is favoured, or growth of nuclei even occurs (e.g., octacalcium phosphate) in equilibrium reactions through progressive accumulation of ordered crystal layers.

1. Structural evidences

Less crystalline reaction products ("paracrystalline"), more amorphous in nature, are formed if aggregation of free reactive particles in the gaseous or liquid phases occurs very rapidly, not through a reversible consolidation equilibrium. Valencies are not ordered according to the structure of a stable lattice, but are satisfied by the nearest available valency sites. This random ordering, involving stretched bonds, distorted valency angles, etc., is thermodynamically unstable; recrystallization will take place when the activation energy reaches the level required for shifting positions to those of the ordered lattice.

Formations of fine crystalline or colloidal precipitates of apatites requires the atomic relationship of the cations (Ca...) and anions ($PO_4$...) to be smaller than for the ideal structure Ca:P = 5:3 = 1.667; it lies more often between 1.3 and 1.6.

The following basic possibilities exist for lowering the Ca/P ratio: (1) separation and removal of calcium ions from single lattice positions; (2) relative increase of the phosphate components (by adsorption of phosphate ions); (3) partial structural change (mixed crystals with apatite-like, but calcium-poor structure type, e.g., octacalcium-phosphate).

To lower the Ca/P ratio from 1.67 to 1.5 according to (2) above by adsorption of phosphate groups at the crystal interface, it is necessary to have a free apatite surface of 150-200$m^3$/g, corresponding to an edge length of 50-100 Å.

Particles of such small dimensions occur in precipitates obtained synthetically, and in biological mineralizations (dentin, bones). In apatitic particles of this size 30% of the phosphate groups are situated on the surface. This is also the proportion expected by physical chemical theory, which does not distinguish between the surface adsorption and the crystalchemical theories of the nature of defect sizes within the structure (e.g. calcium gaps). Thus one and the same process, leading in the macrodomains to highly differential properties, can be encompassed by two different model concepts.

The bigger the particles the more the adsorption concept must be discounted; the explanation of the calcium deficit of apatite crystals that do not obtain the theoretical Ca/P ratio must be sought on a crystalchemical basis, as suggested in (1).

X-ray investigations of precipitated apatite reveal primarily that regular interstice

positions are absent in certain lattices (as superstructure). Their presence would be indicated by additional intermediary reflections in the X-ray diagram. This is not the case in precipitated or bony apatite.
The less crystalline apatite preparations provide an example of paracrystalline crystal structure, which justifies naming such precipitant products "sparingly soluble phosphates with paracrystalline apatite structure". The lattice of such substances is an apatite lattice whose individual unit cells differ from the ideal structure. If one sums up all the elementary cells to obtain the crystal composition, one gets a slightly smaller Ca/P ratio. In agreement with these results the specific gravity of the nonstoichiometric calcium phosphates is smaller than the theoretical value for pure apatite (3.05...3.1 compared with 3.15). The "gap positions" in the crystal lattice must be saturated by uptake of hydrogen ions to satisfy the valencies. These ions, however, are not present as such; rather they react with phosphate and hydroxyl groups to form hydrogen bonds ($O_3PO$-H...$OPO_3$, $O_3PO$-H...OH). These compounds can be verified by their infrared absorption: characteristically for calcium phosphates with Ca/P 1.67 one observes a constriction of the OH frequency at 3600 $cm^{-1}$ and the $H_2O$ bonds at 3420 $cm^{-1}$ and the deformation frequencies at 1635 $cm^{-1}$, also a longer flatter decrease in the absorption curves at 4000 $cm^{-1}$.
The crystallite size of apatite with statistical hydrogen bonding of this kind remain small because of the resulting bond stretching and lattice tension. Substituting foreign ions can stabilize the crystal lattice of nonstoichiometric apatites and remove the lattice tensions. If, for example, Na ions enter for the partial valency saturation, they can substantially represent for $Ca^{2+}$ ($Ca^{2+}$ = 0.99 Å; $Na^+$ = 0.97 Å), so that bond stretching remains insignificant.
In this context of crystal nucleation we should also consider the efficacy of crystallization inhibitors, or potentiating factors (e.g., epitaxy of matrices) as influencing the activation energy. Although the above mentioned precipitation phenomena resemble the biological reactions of inorganic components, in quantitative determinations of the relationship of various factors it is always necessary to study defined surfaces of single crystals.
In summary of the nucleating effects, the precipitated calcium phosphates can be described exactly only by the criteria giving the characteristics of colloidal single particles (size, number, charge); data from methodology using summations over all the sets of observations - average values (such as quantitative analyses, determination of water content, density, etc.) - should be used only as additional evidence. The sparingly soluble calcium phosphates are of particular interest as components of the vertebrate skeleton hard tissues. Their crystallographic properties fulfill important functions concerning also the sequence of events leading to the mineralization of bone. The various steps comprise the precipitation of an amorphous material, the appearance of an initial and a final crystalline precipitate. The initial crystalline precipitate is a highly Ca-deficient apatite; the final precipitate is hydroxyapatite.

## 2. Textural evidences

In bone tissue, wide variations exist in the proportions of organic to mineral material, leading to a wide range of specific gravitities which for diaphyseal bone extends from 1.65 to 2.25 g/cm$^3$.
Results of X-ray diffraction as applied to specific gravity fractions obtained from the diaphyseal bone samples (powder diagrams following Debye-Scherrer) before heating, for all fractions a definite though diffuse pattern is observed; the lines can be attributed to an apatitic crystalline material. The distinct feature of the series is the fact that the lines are clearly more intense for fractions 1.75 and 1.85 than for other fractions.
After heating the material for 1 hr at 850$^{\circ}$C (see below) one observes in fraction 1.65 only lines corresponding to anhydrous tricalcium phosphate ($\beta$-TCP). In fraction 1.75, apatite pattern begins to appear; both the $\beta$-TCP and apatitic pattern are superimposed. As the specific gravity increases, the intensity of the $\beta$-TCP lines progressively decreases whereas the apatitic pattern becomes more and more clearly apparent. In the 2.25 fraction, the $\beta$-TCP lines, though present, are seen only with difficulty.
The integral breath of the (002) peak decreases from the 1.65 fraction up to the 1.95 fraction, from where it stays practically constant. This indicates an improvement of the mean size and/or perfection of the bone apatitic crystals up to fraction 1.95 where the apparent particle size in the c-axis direction reaches about 21.5 nm. The analysis of the pure diffraction profile of the (002) peak indicate that, whereas there is a progressive but small improvement of the crystal size and/or perfection in the c-axis direction, it is mainly an increase in the number of crystals present which accompanies the increase in the mineral load, as the specific gravity goes from 1.65 to 2.25.
Contrary to the amorphous phase , the amount of crystalline material present is to describe by a discontinuous function. This would suggest that the appearance of the crystalline material does not correspond to a single and progressive process and that two steps should be differentiated: a rapid and temporary increase in the number of crystals, followed by a slower and permanent second generation. Relationships that appear to exist between the various materials distinguish two fractions: the first one, non-crystalline and the second, crystalline. In the non-crystalline, the precipitation of the amorphous material takes place at the expense of a liquid phase, in as yet undefined relation with the serum. As the initial crystalline precipitate appears in the low specific gravity fraction, the mass of amorphous material decreases. One may also assume that a transformation takes place, the initial crystalline precipitate being the source of the final crystalline precipitate. Further on, both the precipitation of amorphous material and its subsequent crystallization proceed at about the

same rate. The slow crystallization of the final crystalline precipitate, even though it certainly could as well take place directly. It may be difficult to establish exact composition and structure of the microcrystalline precipitates, they must be used as remote approximation of reality, even when structural concepts are introduced in order to define concepts as voids, surface or internal defects or substitutions.

One has to add to the system ions such as carbonate (in various forms), mucopolysaccharides and collagens, and the various minor elements present in bone ($Mg^{2+}$, $Na^{+}$, $K^{+}$, etc.). In the 1.65 fraction, the two diffusion halos in the Debye-Scherrer-diagram are probably due to the presence of organic material and in particular to the debris collected in that first separated fraction.

Within the sequence of events which lead to the mineralization of structural units of bone, the following steps are to describe:

a) the mineral phase of bone is precipitated primarily as an amorphous material (primary gel);
b) the amorphous component, in a first step, gives rise to an initial crystalline precipitate;
c) the initial crystalline precipitate disappears, being transformed in a following step, into a final crystalline precipitate formed;
d) once this final crystalline precipitate formed, crystallization proceeds at the expense of both the amorphous material and the liquid phase;
e) the apparent crystal size does not change much with time (from 15 to 21,5 nm), and the increase in the mineral load is mainly due to an increase in the number of crystals;
f) the initial crystalline precipitate is an apatite with a composition close to that of OCP;
g) the final crystalline precipitate is an apatite similar to HA;
h) the carbonate is present in more than one chemical form and/or environment, the proportion of which changes with time.

Taken alltogether,these data lead to consider the initial crystalline precipitate to be best described as a calcium-deficient apatitic material, with a composition close to that of OCP; the final crystalline precipitate may be considered as HA. Both crystalline phases would exist in varying proportions in the intermediate specific gravity fractions. In bone, mixtures exist in varying proportions of the two crystalline precipitates rather than assuming the existence of a progressively varying composition of the crystals.

## 3. Transformation/epitaxy pattern for biological species

Recent informations concerning factors controlling the structure of chemical components in the calcium phosphate system confirm previous assumptions based on electron microscopic evidence that the nature of calcified components in biological hard tissues differs crystalchemically. Although the result of X-ray and electron diffraction and micro-morphological studies clearly show that, for example, intact dental enamel is of an apatite type, the same methods show deviations from this crystalline structure in poorly mineralized and altered hard tissues.

The solubility difference between crystalline apatite and amorphous calcium phosphate gels permits the (re)crystallization of bone crystals. The phenomenon of an increase in particle size is a general physicochemical rule. Single crystal formation, however, is only possible from monodispersed solutions: only simple ions can enter the lattice positions during successive build up of the crystal lattice planes. The simple ions ($Ca^{2+}$, phosphate, etc.) are present in low concentration, corresponding to the difference in solubility between the amorphous and crystalline phases or the dissociation of organic phosphate complexes. The mobility of ions, which is defined for ideal aqueous solutions, is much smaller in gels. The small concentration and the low mobility of ions contributing to the crystallization centers necessitates a slow reaction rate, thereby permitting an ordered separation of the crystalline phase. The transformation of the amorphous material of the primary gel into an ordered crystal occurs in general at the same rate in bone, dentine and enamel. The collagen fibers of bone and dentin matrices, besides their mechanical influence on orientation also have an epitactic effect. This function is, however, not as extensive and ideal as in the case of the keratin matrix of dental enamel; the degree of crystallinity in dentin and bone is therefore markedly less. The regularity of enamel crystals cannot then be solely explained by analogy with the process of apatite crystal growth on a mineralogical and geochemical basis.

From crystalchemical and in vitro reactions studies, it appears that crystal growth under physiological conditions takes place on an apatite-like phosphate-rich precursor, including overgrowth processes on the protein matrix of the hard tissue. The primary lattice order corresponds to elementary units of octacalcium phosphate, whose immediate transformation to the apatite coordination excludes the formation of amorphous clusters and is responsible for the growth in length of enamel crystals. This consolidation reaction is markedly influenced by accompanying and trace ions (fluoride, carbonate, magnesium). In pure aqueous solutions of calcium and phosphate ions at acid pH, calcium hydrogen phosphates $CaHPO_4\ 2H_2O$ would be stable at supersaturation. At weakly acid pH at lower temperatures - i.e. physiological conditions - octacalcium phosphate, $Ca_4H(PO_4)_3 \cdot 2H_2O$ would be the stable form. Octacalcium phosphate can only be obtained by exact adherence to the conditions of stoichiometry. Hydrolysis occurs toward the neutral and alkaline pH region, and the previously mentioned transformation takes place, here under crystalchemical aspects: preformed apatite-like lattice regions are progressively transformed to an extensive apatite-type (in part paracrystalline in character) along the crystallographic orientation $\{100\}$ of the octacalciumphosphate lattice. The composition of these metamorphic products, i.e., their molar relationship Ca/P varies in the limits 1.33...1.67 (corresponding to octacalciumphosphate and apatite, respectively). The formed compound corresponds to the mineral type of the particular biological hard tissue.

The transformation is greatly influenced by traces of fluoride - also by alkaline pH:

$$Ca_4H(PO_4)_3 \cdot 2H_2O \text{ --- } (H_2O,\ F^-) \text{ ---}\rightarrow Ca_5(PO_4)_3 \cdot (OH,F)$$

The determining effect of therapeutic fluoridation doses in trace concentrations is probably the induction step of the apatite structure to the mineral portion of hard tissues e.g. enamel; this is more effective than just the solubility difference between hydroxyapatite and fluorapatite, standing in discussion for a long period.

The carbonate content of bone and tooth hard tissue also can be incorporated into the para-crystalline reaction scheme of the OCP/HA metamorphosis without great effort: partial substitution of $HPO_4$ groups of octacalcium phosphate gives a reaction product which in its basicity tends toward the apatite type:

$$Ca_4(PO_4)_2(HPO_4)_{0,4}(CO_3)_{0,6}$$

The morphological properties of this compound correspond particularly to the mineral of bones and dentin.

## Synergetic/antagonistic functions of trace elements

In the simultaneous effects of several foreign components, even of trace elements, one must consider synergism (e.g. magnesium/carbonate) and antagonism (fluoride/carbonate). The actual course of biological calcification is determined by the concentration relationships in the secondary components. Summarizing, and introducing the trace elements aspects, bone composition is given by a carbonate-containing, apatitic calcium phosphate with a disturbed - paracrystalline - lattice containing also amorphous and adsorbed phases. The composition may as well be interpreted as the intermediate structure interlayered between hydroxyapatite and octacalcium phosphate, both substituted by Na, Mg, F, $CO_3$, $HPO_4$, and a series of other substituents. Investigations with radioactive tracers showed that this series may be deposited with and again mobilized from skeletal bones; hard tissues offer homoeostatic functions by storage at excessive supply of trace elements and by an output at their need. Compared with other tissues of the human body it follows from the estimated pool size in bone that a high level of trace elements is available there. Informations on the disposition and mobilization of the pool are obtained by the measurement of element concentrations as function of factors like stress, change of hormone concentrations etc. Any element with the exception of selenium shows immediately after stress propagation an increase in concentration. The raise is significant with $P<0.001$ in zinc and iron, and with $P<0.05$ in cobalt and rubidium. The level of zinc remains still for two hours significantly elevated. Alteration of metabolic rates is observed in pregnancy for copper, zinc, and chromium; also the metabolism in selenium is influenced by the sexual hormones.

The methods for microanalyses of the elements at the nanogram and subnanogram level, by neutron activation and flameless atomic absorption spectroscopy, are still not yielding standard routines, even determining average values of the elements in the samples but without evidence of the chemical bonding state. Typically trace elements in the tissues are bound to organic molecules, mostly proteins. Therefore, although knowledge of the changes caused by a companion ion gives some insight into the specific reaction mechanism, it is still necessary to consider that in the presence of several foreign components these effects may be enhanced or inhibited, be it through prevention of nucleation or crystal growth, or through substitution or structural change, which will be revealed by altered dissolution rates. These factors will be discussed in greater detail in relation to crystallization of apatite.

## 1. Effects of anions

The reaction mechanism of the fluoride ion in the crystallization of apatite in physiological milieu may be derived from lattice analogies of apatite and octacalciumphosphate.

Under the same reaction conditions in which it would be stable in a fluoride-free medium, octacalcium phosphate reacts in the presence of very small amounts of fluoride (solutions containing 10-100µg $F^-/l$) homogeneously forming fluorhydroxyapatite; in comparison, crystalline octacalcium phosphate reacts heterogeneously to form apatite. The reaction mechanism of fluoride actions can be elucidated unter two headings:

(1) Formation of fluorapatite nuclei along the octacalcium phosphate layers in the apatite-like subcells: the isomorphous hydroxyl and fluoride are able to replace each other completely. If fluoride occupies a corner position in an apatitic subcell of octacalcium phosphate, then the layer succession must stop there, since the hyroxyl groups at the apatitic corner positions are required as proton acceptors.

(2) Growth of fluorapatite nuclei by epitactic growth of hydroxyapatite. The leaf-form habitus of the formed crystallites is determined by this mechanism.

Carbonate ions appreciably interfere with the crystallization of apatite from homogeneous solution: surface formation of the crystal and also particle growth itself is already inhibited by small concentrations of carbonates; at higher concentrations, amorphous calcium phosphate appears. Under certain circumstances, carbonate phosphate can crystallize, even under biological conditions, namely through reaction between $CaHPO_4 \cdot 2H_2O$ and $CaCO_3$ in aqueous suspension. The action of calcium carbonate, as a factor reducing the crystallinity of biological calcium phosphate, is here a general one: citrate and pyrophosphate reduce crystallinity to an even greater extent.

## 2. Effects of cations

The divalent ions of magnesium, manganese, and iron can affect the formation of the whitlockite structure in pure calcium solutions under conditions in which octacalcium phosphate or hydroxyapatite appear. The reaction mechanism of this change is explained by the induction of the whitlockite structure by foreign components, that is, the different reaction medium provides the proper conditions for the formation of the new phase. In the pure calcium phosphate system, at normal temperature, the energy required for nucleation is higher for whitlockite, corresponding to the size and electrical charge relationship of the ions; hence hydroxyapatite is stable. By introduction of the foreign ions - divalent metal ions - into the bonding field the nucleation energy for whitlockite is lowered to the level at which this structure becomes stable at normal temperatures (at higher temperatures the reaction is also shifted to the direction of the whitlockite structure, without foreign substitution).

Size and charge relations alone are not sufficient tools to characterize the ions that can affect structural change. The ionic radii of $Mg^{2+}$ (0.66 Å), $Fe^{2+}$ (0.74 Å), and $Mn^{2+}$ (0.80 Å) hardly differ from those ions that are ineffective, such as $Ni^{2+}$ (0.69 Å), $Co^{2+}$ (0.72 Å), $Zn^{2+}$ (0.74 Å), $Cd^{2+}$ (0.97 Å). It follows, therefore, that only slightly polarized divalent ions can function as nucleators for the tertiary phosphates like whitlockite ($\beta$-tricalcium phosphate). An analogous differentiation is shown by the instability of the divalent ions of Mg, Mn, and Fe, in forming stable ammine complexes within water, in contrast of $Zn^{2+}$, $Co^{2+}$, $Ni^{2+}$, and $Cd^{2+}$.

Nucleation as described above leads to formation of mixed crystals of $\beta$-tricalcium phosphate containing foreign ions. The further possibility that structure induction of this kind precedes nucleation has been demonstrated in experiments with freshly precipitated trimagnesium phosphate. Among the cations known to be a factor influencing only the degree of crystallinity, $Mg^{2+}$, $Sr^{2+}$, and especially $Cu^{2+}$ (effective even at $10^{-6}$ moles) inhibit the crystallization of octacalcium phosphate.

The simultaneous action of several foreign ions may result in synergism or antagonism.

Fluorine alone improves the calcification of hard tissue, even at low calcium and phosphate concentrations, the effect being proportional to the dose at low levels; with higher doses the effect is enhanced. Even at high calcium and phosphate concentrations the synergistic effect of fluoride is unmistakable, even though the individual reactants may not be relevant for mineralization. The basis of this synergistic effect is reaction steering, which evens out fluctuation in phosphate concentration. Crystallization of calcium fluoride (Fluorite) at extreme fluoride concentration is hereby avoided. Fluoride and carbonate show a distinct antagonism: increasing the fluoride concentration of biological hard tissues decreases its carbonate content. This explains why fluoride improves the crystallinity of apatite; since only small concentrations of carbonate can exist in the lattice of well-crystallized apatite. Besides inducing the whitlockite structure in the calcium phosphate system, magnesium shows additional synergistc effects: it enhances the inhibition of carbonate on crystallization of apatite and also stabilizes the amorphous calcium carbonate phosphate phase.

The reaction mechanism of molybdenum (molybdates) is synergistic like fluoride; the fluoride concentration in the hard tissue is not altered by this reaction. However, fluoride uptake by subsurface layers is improved in the presence of chloride. Also the mutual interaction of chloride and fluoride on retention has been noted. Vanadium, strontium, and lead as trace elements are known to react by isomorphous substitution in the apatite lattice under environmental conditions. By this means vanadium causes a definite increase in the length of the crystals. Different crystal types have also been reported for iron and manganese in the presence of fluoride; presumably they are present as fluorocomplexes.

This crystalchemical explanation is furthermore applicable to the synergistic effect of aluminium and tin, whereas the striking antagonistic action of selenium cannot be explained thus, but may be instead understood as an inhibition of mineralization by changes in the protein components of hard tissues. Finally, an interesting antagonistic action of beryllium should be noted: inhibition of the formation of whitlockite crystals, an effect aided by magnesium.

## High-temperature reactions of calcium phosphates

Chemical stability or disintegration is reflected also in the high temperature behaviour of hydroxyapatite. It is known that in the temperature field above 1000$^{o}$C hydroxyapatites are no longer stable. However, the change is not a simple and homogeneous one; furthermore, the reactions are strongly dependent on secondary conditions of the synthesis of these crystals or the formation of them.

Stable phases in the high temperature region of the calcium phosphates, according to the constitution diagram of the system $CaO/P_2O_5$ (if one disregards the polyphosphates) are the mono-, di-, tri-, and tetraphosphates. Two modifications each of the first three named compounds are known.

| | $CaP_2O_6$ | $Ca_2P_2O_7$ | $Ca_3P_2O_8$ | $Ca_4P_2O_9$ |
|---|---|---|---|---|
| F.P. ($^oC$) | 985 | 1335 | 1720 | 1630 (incongr.) |
| Crystal class | monoclin(β) | orthoromb.(α)<br>tetragonal(β) | orthoromb.(α)<br>tetragonal(β) | monoclinic |

The results of the high temperature firing experiments can be placed in the phase diagram. In fact, in the temperature field from 700 to 1000$^oC$, the precipitated colloidal apatite (which contains $H_2O$ and $HPO_4^{2-}$) is already very similar to the hydroxyl-free tricalcium phosphate stoichiometry. This is because the Ca/P ratio in $Ca_{10}(PO_4)_6(OH)_2$ $+HPO_4^{2-}$ appears reduced, and approaches stoichiometrically that of $Ca_3(PO_4)_2$ like:

$$10/6 = 1.67 \longrightarrow 1.50 = 3/2.$$

Accordingly the reactivity of this nonstoichiometric, finely divided compound, especially the thermal degradation and dehydration will be favoured, so that the hydroxy groups of the precipitated form of apatite combine with protons of the absorbed $HPO_4^{2-}$.

At 1000$^oC$, the phase plot of the precipitated product corresponds generally to that of the pure water-free calcium phosphate:

$$Ca_{10}(PO_4)_6(OH) + \text{adsorb. } HPO_4^{2-} \text{ --- 700 to 1000}^oC \longrightarrow 3\beta Ca_3P_2O_8 + H_2O,$$

with some $Ca^{2+}$ position present in excess. This means lattice imperfections or residual colloidal apatite absorbed. However, these faults cannot be detected in the X-ray diagram because of line broadening. If the astoichiometric compounds are present in fairly large amounts, then on higher temperature rise of the material above 1400$^oC$ a further change will take place, namely on the tetracalcium phosphate

$$2Ca_3P_2O_8 + Ca_{10}(PO_4)_6(OH)_2 \text{ --- 1400}^oC \longrightarrow 4Ca_4P_2O_9 + H_2O.$$

$\beta$-$Ca_3P_2O_8$ alone would transfer at 1400$^oC$ to $\alpha$-$Ca_3P_2O_8$, and this is shown as the reaction path of the main portion of the precipitated apatite at high firing.

Pure (e.g. hydrothermally grown) hydroxyapatite retain at heating to temperatures above 1250$^oC$ the permanent apatite X-ray diffraction diagram. At temperatures above 1450$^oC$ the $\alpha$-$Ca_3P_2O_8$-phase appears, with additional lines which correspond again to the $Ca_4P_2O_9$. Thus the thermal decomposition of the hydroxyapatite takes place according to

$$Ca_{10}(PO_4)_6(OH)_2 \text{ --- 1450}^oC \longrightarrow 2\,\alpha\, Ca_3P_2O_8 + Ca_4P_2O_9 + H_2O.$$

Using microprobe analysis to investigate Ca/P stoichiometry produced inconsistent information. Since the thermal transformation of the solid state reaction takes place without mass transport, the products remain at the reaction site, and even in regions less than 1$\mu m^3$ only an average composition value for the transformed microcrystalline composite mass will be obtained.

Fluoridated apatites are affiliated closely to the reaction scheme of the hydroxyapatites in as much as these are loosing their fluorine contents at moderate firing. Fluorine loss occurs especially because of the simultaneous presence of $HPO_4^{2-}$, resp.

pyrophosphate in the materials with the beginning thermal treatment. The pyrophosphate reacts with fluoride further to volatile $POF_3$ compounds.
Stoichiometric pure fluorapatite $Ca_{10}(PO_4)_6F$ on the other hand, is thermally more stable than hydroxyapatite and keeps at temperatures as high as 1500°C and more. The reactions and transformations in burnt bones are to be demonstrated schematically by their thermal behaviour up to 700°C (critical temperature). Since bone is composed of both organic and inorganic matter, fresh long bone, for example, when it is dried, shows weight loss of about 35%. This is mainly due to the loss of water and protein. Fat, carbohydrates etc. together amount to less than 10% of this value. The ratio of total calcium to phosphorus (P) has been determined as 1.70. The calcium carbonate content reduces the Ca/P ratio of bone mineral to a true value below the apatite stoichiometry (1.67), implicating a gradual thermal decomposition of bone, above all by the loss of the adsorbed and structural water (100 to 400°C) and by a recrystallization mode. From 600°C onwards the split of the carbon dioxide follows.
From the increasing firing temperatures proves a significant decrease of the mechanical strength reaching a minimum between 400 and 600°C. Then strength of the bone rises again with temperature, to a high maximum at 800°C. Micromorphological investigations verify the transition, from the lamellar bone texture into a homogenuous texture, which is to be seen also by its cretaceous appearence. This critical strengthening behaviour is connected with sintering processes of the bone mineral in the range of 700 to 1000°C, by reaction of intermediate pyrophosphate phases with the primary apatite crystallites into whitlockite as shown by formula on page 14 . The coalescence of the bone mineral crystals gives rise to a volume loss, which is the typical shrinkage of burnt bones.
Disruptive reaction conditions with thermal reactions in the calcium phosphate system yielding pyrophosphate at 600 to 800°C are also known in the presence of other salts like $(NH_4)_2HPO_4$ and $NH_4NO_3$, as well as recombination of calcium pyrophosphate and hydroxyapatite into tricalcium phosphate.

## References


Behne D (1975) Untersuchungen zu Speicherung, Mobilisierung und Transport von Spurenelementen im menschlichen und tierischen Organismus. In: Bethke K, Bidlingmaier F (eds) Spurenelemente in der Entwicklung von Mensch und Tier. Urban & Schwarzenberg München. pp 77-87.

Dahlem Workshop on Biological Mineralization. Mechanisms of Mineralization. Springer Berlin Heidelberg New York (1984)

Frondel C (1943) Mineralogy of the calcium phosphate in insular phosphate rocks. Amer Mineral 28: 215.

Gruner JW, McConnell D (1937) The problem of carbonate apatites. Z Krist 97: 208.

Hayek E, Newesely H, Hassenteufel W, Krismer B (1960) Zur Bildungsweise und Morphologie der schwerlöslichen Calciumphosphate. Monatshefte F Chemie 91: 249-262.

Haynes G (1985) Age profiles of elephant and mammoth bone assemblages. Quaternary Res 24: 333-345.

Herrmann B (1976) Neue Ergebnisse zur Beurteilung menschlicher Brandknochen. Z Rechtsmed 77: 191.

Herrmann B, Newesely H (1982) Dekompositionsvorgänge des Knochens unter langer Liegezeit. I. Die mineralische Phase. Anthrop Anz 40: 19-31.

Klement R (1938) Die anorganische Skeletsubstanz, ihre Zusammensetzung, natürliche und künstliche Bildung. Die Naturwissensch 26: 145.

Matheja J, Degens ET (1969) The role of phosphates in biochemical reactions. N Jahrb Mineral, Abh 110: 227-251.

McConnell D (1952) The problem of carbonate apatites. Bull soc franc minéral et crist 75: 428.

Newesely H (1966) Kristallchemische und mikromorphologische Untersuchungen der schwerlöslichen Calciumphosphate. Fortschr chem Forsch 5: 688-746.

Newesely H (1970) Die mineralogisch-geochemische und die biogene Kristallisation des Apatits. Biomineralisation (Forsch-Ber Akad Wiss Mainz, math-naturwiss Klasse) 2: 38-47.

Newesely H (1970) Factors controlling apatite crystallization with particular reference to the effect of fluoride and accompanying ions. Adv oral biol 4: 11-42.

Newesely H (1972) Mechanisms and Action of Trace Elements in the Mineralization of Hard Tissues. Zyma Lyon.

Newesely H (1977) Biokristallographische Aspekte in der Archäometrie. Arch Naturwiss (Mainz) 1: 81-84.

Newesely H (1977) High temperature behaviour of hydroxy- and fluorapatite. J Oral Rehabil 4: 97-104.

Newesely H, Herrmann B (1980) Ab- und Umbauvorgänge der biologischen Hartgewebe (Knochen,Zähne) unter langer Liegezeit. Berl Beitr Archäometrie 5: 175-186.

Newesely H, Hosemann R, Uther B (1980) Kollagen-Einschlußverbindungen. Z Naturforsch 35c: 177-187.

Quinaux N, Richelle LJ (1967) X-ray diffraction and infrared analysis of bone specific gravity fraction in the growing rat. Israel J med sci 3 (5): 677-690.

Strunz H (1970) Mineralogische Tabellen. 5th edition. Akad Verlagsges Leipzig.

# METHODICAL ASPECTS OF THE DETERMINATION OF Pb, Cd AND Mn IN FOETAL BONE

M. Hedrich, U. Rösick , P. Brätter
Hahn-Meitner Institut Berlin GmbH
Trace Elements in Health and Nutrition
Glienicker Str. 100
1000 Berlin 39, FRG

R.L. Bergmann and K.E. Bergmann
Bundesgesundheitsamt Berlin
Institute of Social Medicine and Epidemiology
General-Pape Str. 62
1000 Berlin 42, FRG

## Summary

Zeeman compensated graphite furnace atomic absorption spectrometry was used to determine the contents of Pb, Cd and Mn in wet digested rib bones of foetusses and children up to 2 years of age. The method was scrutinized with respect to wavelenght, acid and sample concentrations. The accuracy of the measurements was clearly influenced by lot, individual quality and age of the graphite tubes. Monetite ($CaHPO_4$) of high purity was used to supervise contamination and quality control was reassured by H5 animal bone, a standard reference material of the International Atomic Energy Agency. Examination of so far 66 bone samples resulted 0.5-4 µg Pb, 7-115 ng Cd, and 0.1-5 µg Mn per gram of defatted dry matter.

## Introduction

In preparation of an epidemiological study concerning intake, fixation and remobilization of bone seeking essential and toxical trace elements in the foetal and infant skeleton it was to be tested, if a direct determination of Pb, Cd and Mn in bone samples with graphite furnace atomic absorption spectrometry (GFAAS) was possible with the necessary precision and accuracy. It is well known that a direct atomization of Pb and Cd in the presence of calcium phosphate is problematic because of unexplained signal depressions (1-2). In spite of this it has been tried increasingly during the past years to determine lead in bones and teeth directly to avoid element separation prior to analysis, which is time consuming and susceptible to contamination (2-6). A scrutinization of these analytical conditions seemed to be needed urgently, however, as after an interlaboratory comparison Stack and Delves (7) had found unacceptable differences in the results of tooth lead determinations between the laboratories involved.

## Experimental

As sampling material we received rib bones of foetusses from abortions after the 4$^{th}$ month of pregnancy and of children who have suffered from sudden infant death. The samples are taken with a normal scalpel from the 6$^{th}$ rib bone on the left hand side in amounts between 10 mg and 1g and preserved at -18$^{o}$C. Larger samples are cut down by a low speed Buehler-Met diamond saw with 70% ethanol as cooling agent. Diethylether is used to defat the bones in quartz tubes for at least 6 hours in a Soxhlet reflux extraction apparatus. After drying in air over night the bone samples are dissolved in concentrated nitric acid and wet digested for a period of 9 hours at 160$^{o}$C. Merck's nitric acid of suprapure quality contains up to 5 µg of each Pb and Cd per liter. As this is within the expected range of the Cd content of bone, we distilled the $HNO_3$ in quartz vessels prior to use. 1 ml of concentrated $HNO_3$ is used to dissolve each bone sample from 10 mg to about 140 mg of weight. Wet digestion is performed in quartz tubes covered with quartz lids enclosed in 2 teflon coated Seif bombs (8) to allow 12 samples to be digested under the same conditions at the same time. The solutions obtained are diluted with doubly distilled water to a volume of either 5 or 10 ml and analyzed without further preparation.

A Perkin-Elmer spectrometer 5000 with Zeeman compensation and automatic sampler AS 40 was available to us for these examinations. The injected sample volume was 16 µl in every case. The evaluation program for peak area integration has been developed in our laboratory (9). Details of the measuring conditions are compiled in table 1.

Table 1: Instrumental Parameters in the Determination of Pb, Cd and Mn with GFAAS

| Element | (nm) | Lamp | Ashing/Atomization ($^{o}$C) | Plat-form | Standardaddition (µg/l) | | |
|---|---|---|---|---|---|---|---|
| Pb | 283.3 | EDL/HKL | 700/2000/MPH*/Gasstop | + | 5 | 10 | 15 |
| Cd | 228.8 | EDL | 600/1700/MPH/Gasstop | + | 0.5 | 1 | 1.5 |
| Mn | 279.5 | HKL | 1100/2200/1s/Gasstop | - | 1 | 2 | 3 |

*MPH = maximum power heating

For Pb and Cd the conditions of the STPF (stabilized temperature platform furnace) concept are largely considered. Because of the high phosphate concentration of the bone matrix the given ashing temperatures could be used without loss of elements and modifying the matrix was not necessary. The reason for not atomizing Mn with maximum power heating was a signal artefact that appeared when hollow cathode lamps (HKL) were used and that was not eliminated at the beginning of the examinations (9).

Animal bone powder H5, a standard reference material of the International Atomic Energy Agency (IAEA, Vienna), was used to control the quality of the analyses and a self synthesized monetite ($CaHPO_4$) with very low contents of Cd and Pb served as contamination supervision of our pressurized digestions.

## Results and Discussion

At first we followed the question, if the most sensitive resonance line of Pb at 217.0 nm or the less sensitive one at 283.3 nm was the most suitable for our bone analyses. In these preexaminations we used hydroxyapatite (Merck, art. no. 2143) as an in acids readily soluble bone analogous matrix. The substance contains about 1 µg Pb/g. Figure 1 shows the lead signals of both wavelengths after atomization from a nitric acid solution following the conditions of table 1.

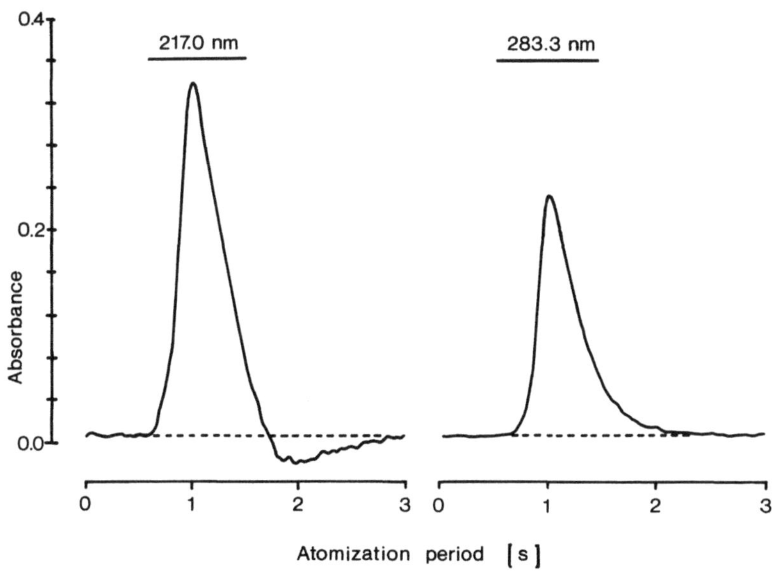


Figure 1: Pb signals at 217.0 nm and 283.3 nm after atomization from a solution of hydroxyapatite in 6.5% (v/v) $HNO_3$ (measurement with Zeeman compensation).

The measurement at 217.0 nm exhibits an uncompensated oscillation below the baseline despite Zeeman operation and STPF conditions. Peak area integration reveals incorrect results in this case and we assume that this holds true for peak height evaluation as well. This phenomenon cannot be explained by failure of the Zeeman compensation (9) and it does not appear when lead standards are analysed in nitric acid solution. In comparison to this measurement at 283.3 nm deliver distinct signals that are evaluable without any problems. So we used this resonance line exclusively.

The acid used for digestion may lead to signal depression in the determination of Pb (10). Table 2 informs about the results of testing this possible interference.

Table 2: Influence of the $HNO_3$ Concentrations in the Digested Solutions on the Pb Signals

| $HNO_3$ Concentration %(v/v) | Peak Areas of the Pb Signals (mAbs·s) H5* | LA1* |
|---|---|---|
| 3.25 | 34.5 ± 0.3 | 97.1 ± 0.8 |
| 9.75 | 36.0 ± 0.6 | 93.4 ± 1.0 |
| 16.25 | 35.2 ± 0.5 | 93.8 ± 1.0 |
| 22.75 | 35.2 ± 1.2 | 93.7 ± 0.7 |

* bone mass 2.6 mg/ml

As the results in the table show the peak area of the lead signal is independent of the $HNO_3$ concentration in wide ranges under our conditions of analyses. This is the case for the H5 standard as well as for our internally used animal bone standard LA1 with a higher content of Pb. Thus samples with high trace element contents could simply be diluted with distilled water to reach calibration linearity.

After finishing these preexaminations production of analytical results was started. So far 8 series of analyses have been measured containing 12 samples each. Each 12 samples were wet digested under the same conditions at the same time and consisted of 9 to 10 bones, one H5 standard, and 1 or 2 samples of monetite. Standard additions were not performed with every sample but merely with monetite, H5 and 1 or 2 bone samples of each series. The linear range of the calibration function was not left in any case. The results of the standard additions - expressed as characteristic masses $m_o$ - are shown in figure 2. For clearness, the different analytical series I thorugh VIII are marked off by dividing lines.

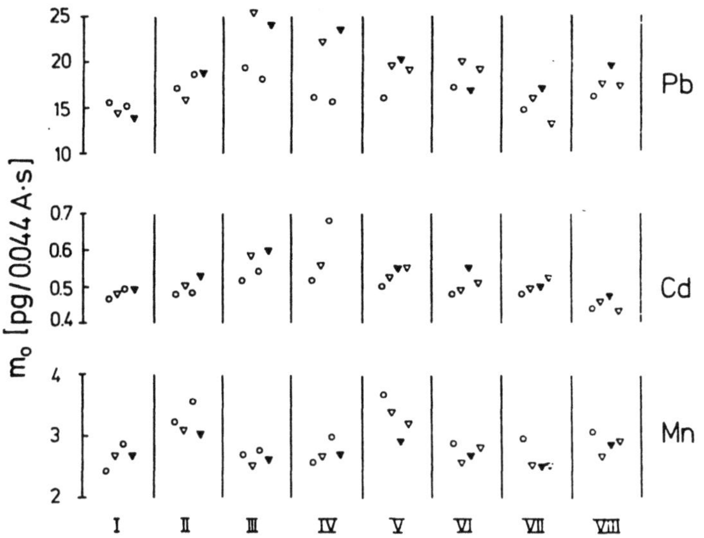


Figure 2: Characteristic masses $m_o$ of Pb, Cd and Mn in samples of bone and $CaHPO_4$ of 8 analytical series. (▽ bone, ▼ H5 standard, o $CaHPO_4$)

Looking at the results for Pb, there is quite a good agreement in the values $m_o$ between the standard added samples of series I and - with limitations - also of series II. In the third and fourth series the bone samples show larger values of $m_o$ as $CaHPO_4$ with differences up to 40%. The following analytical series exhibit a similar behaviour, although it is less marked. Our retrospective search for explanation of this unexpected effect resulted the following:

- starting with series III a new lot of graphite tubes was taken into use,
- the only difference between the examined samples was their content of calcium phosphate. 2 mg of monetite were weighed in per ml of the digested solution, whereas for H5 and the bone samples the concentrations ranged from 5 to 14 mg/ml.

Therefore we checked, if the calcium phosphate concentration of the sample solution influences the slope of the calibration function. Figure 3 demonstrates that there is a dependence indeed.

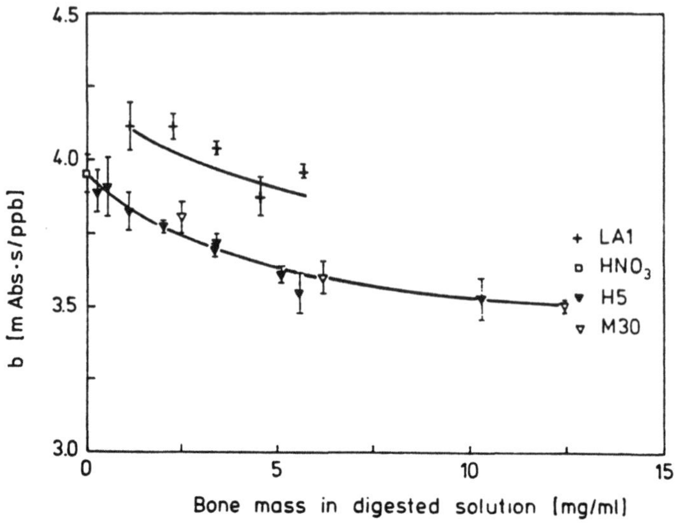


Figure 3: Influence of the bone mass in the digested solution on the slope b of the calibration function in the determination of Pb.

In addition to the calcium phosphate concentration other matrix effects have an influence, too, as the values of the animal bone LA1 do not follow the same curve as the standard H5 and the bone M30 do. Accordingly, correct analytical results can only be obtained, if the standard addition method is applied consequently on every bone sample to be analysed for Pb.

Similar problems as in the determination of lead did not occur in the analysis of cadmium. As can be seen from figure 2, the characteristic masses of $CaHPO_4$ and of the bones have very similar values within the same analytical series. The outlier in series IV is due to a graphite tube that broke shortly after this measurement. Increasing contents of calcium phosphate did not influence the slope of the calibration function in the Cd analyses. Measurements with $CaHPO_4$ concentrations between 1 and 25 mg/ml of the sample solution revealed values of $m_o$ reaching from 0.45 to 0.48 pg.

Determination of Mn was performed without platform and without maximum power heating. In spite of this fact, figure 2 exhibits a good agreement of the values $m_o$ within a series and among the different series. The deviations within the series II and V are due to the large number of measurements this graphite tube had already been used before. This can be avoided, if the tubes are taken out of the analytical process in time.

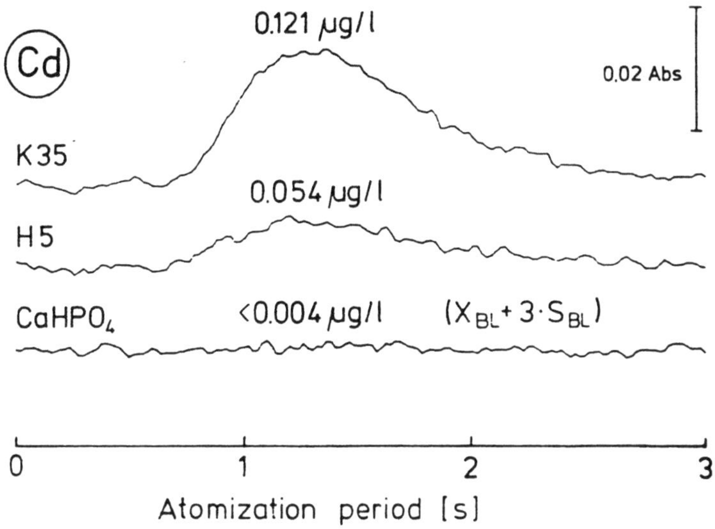


Figure 4: Cd signals of analysical samples after pressurized digestion. K35 is a bone sample.

Considering the analyses of Cd and Pb a strict supervision of contamination was possible by our purified monetite. Figure 4 shows the Cd signal of $CaHPO_4$ after the pressurized digestion and in comparison to this the signals of the standard H5 and a bone sample. Evaluation of the $CaHPO_4$ measurements resulted that less than 0.3 ng of Pb and less than 0.04 ng of Cd were introduced into the sample solution by the digestion procedure. Thus, even concerning determination of Cd in H5 standard, our analyses may be considered free of blank values. We do not have an equivalent possibility of supervision for Mn yet, as the Mn contents of our monetite preparations are about 0.6 µg/g and could not be lowered any further by our cleaning procedures.

Table 3: Results of the Quality Control by SRM H5 (Animal Bone)

| Element | n | Determined | Recommended (11) |
|---|---|---|---|
| Pb | 7 | 3.1 ± 0.3 mg/kg | 2.97 ± 0.26 mg/kg |
| Cd | 8 | 6.0 ± 1.0 µg/kg | 22.9 ± 5.7 µg/kg |
| Mn | 6 | 0.66 ± 0.11 mg/kg | 0.76 ± 0.06 mg/kg |

Results of our quality control with H5 are compiled in table 3. For the elements Pb and Mn there is a good agreement with the certified values (11). Yet the Cd content found by us is far below the recommended value of the IAEA. Because of our effective contamination control, however, we are convinced of the accuracy of our value.

Table 4: Results of Bone Analyses

| Element | This study[1)] | Simon and Liese (12)[2)] |
|---|---|---|
| | n = 66 | n = 65 |
| Pb | 0.5 - 4 µg/g | 3 - 15 µg/g |
| Cd | 7 - 115 ng/g | 4 - 64 ng/g |
| Mn | 0.1 - 5 µg/g | |

[1)] rib bones of foetusses and children, age: -0.5 to 2 years, referred to dry weight
[2)] rib bones of patients with bronchial cancer, age: 35 to 76 years, referred to wet weight

Table 4 shows our analytical results of foetal and infant bones in comparison to the values which Liese and Simon (12) have determined in rib bones of patients from Berlin who have suffered from bronchial cancer. These values do not refer to defatted dry matter of the bones as in our case but rather to wet weight and derive from an age group between 35 and 76 years. Despite the different reference weights in both studies it can be recognized that foetusses and children bear a significantly lower Pb burden compared to the adults. Reasons for this may be the accumulation of Pb in bone during the lifetime and possibly the lead reduction in fuel that may have become effective in the meantime. The Cd contents in bones of children and adults is about the same and for Mn we did not find any informative values for comparison.

## References

(1) Oelschläger W, Lautenschläger W (1977) Fresenius Z Anal Chem 287, 28.

(2) Simon J, Liese T (1983) Fresenius Z Anal Chem 314, 483 .

(3) Langmyhr FJ, Kjuus I (1978) Anal Chim Acta 100, 139.

(4) Grandjean P, Nielsen OV, Shapiro IM (1979) J Environ Pathol and Toxicol 2, 781 .

(5) Ewers U, Brockhaus A, Winneke G, Freier I, Jermann E, Krämer U (1982) Int Arch Occup Environ Health 50, 139.

(6) Drasch G, v Meyer L, Kauert G (1985) in: Lekkas TD (ed): Heavy metals in the Environment , Vol. 1. Athens. p433.

(7) Stack MV, Delves HT (1982) Tooth-lead analysis - an interlaboratory survey. in: Egan H, West TS (eds) IUPAC collaborative interlaboratory studies in chemical analysis. Pergamon Press, Oxford. p115 .

(8) Schramel P, Wolf A, Seif R, Klose BJ (1980) Fresenius Z Anal Chem 302, 62.

(9) Rösick U, Parlow A, Fürstenau G (1986) in: Welz B (ed): Fortschritte in der atomspektrometrischen Spurenanalytik. Verlag Chemie, Weinheim, Vol. 2, p225.

(10) Behne D, Brätter P, Wolters W (1975) Z Anal Chem 277, 355.

(11) IAEA Progress Report No 1: Intercomparison of minor and trace elements in IAEA animal bone (H5) (1982).

(12) Liese T, Simon J (1984) Fresenius Z Anal Chem 317, 482.

# ALTERATION OF CHEMICAL COMPOSITION OF FOSSIL BONES BY SOIL PROCESSES AND GROUNDWATER

C.T. Williams
Department of Mineralogy
British Museum (Natural History)
Cromwell Road
London SW7 5BD
U.K.

## Introduction

Anthropologists have at their disposal several tools to enable the reconstruction of dietary or nutritional habits of ancient populations. One such tool is preserved organic tissue in the form of fossil bone. There are several biological processes that can accumulate some trace elements into living bone, and many studies have attempted to reconstruct the dietary habits or health of ancient populations from an analysis of trace elements in fossil bone. Elements considered to be useful in palaeonutritional studies of this kind include Sr, Zn, Ba (e.g. Sillen and Kavanagh 1982, Klepinger, 1984) and Pb (e.g. Patterson et al. 1987).

An a priori assumption in palaeonutritional studies is that post-mortal exchange of the trace element between the bone and the immediate environment has not occured (or is at best minimal) in order that the primary in vivo concentration is not disturbed by secondary diagenetic processes. However, recent studies have shown that diagenetic alteration of bone during fossilisation involves a wide range of trace elements, including some employed in palaeonutritional studies (e.g. Klepinger et al. 1986), and that the a priori assumption for palaeonutritional studies is often invalid. Interpretation of these data have indicated that instead of providing palaeonutritional information, the trace element content of fossil bone may instaed provide information on aspects of the palaeoenvironment during the diagenetic processes (e.g. Henderson et al. 1983, Williams and Marlow 1987).

There are many interrelated factors involved in the diagenetic alteration of fossil bone. These factors include compositional factors such as groundwater, soil and sediment composition; physico-chemical factors such as Eh, pH and the temperature of the micro- and macro-environment, element transport, uptake and diffusion; physical factors such as bone compaction and crystallinity, and sediment porosity, and biological processes such as the role of microorganisms.

Although it is not possible to account for all these factors, the aims of this paper are to attempt to assess which of them predominate in the alteration of fossil bone from two sites in East Africa and, where possible, to identify the trace elements that are least affected by diagenesis. This paper also summarises some of the results of a larger project involving bone fossilisation processes being undertaken in the Departments of Palaeontology and Mineralogy, British Museum (Natural History). The elements Sc, Ba, La, Ce, Hf, Th and U were selected from a range of 24 elements (see Henderson et al. 1983, for a full list) to illustrate aspects of bone alteration by soil processes and groundwater.

## Samples studied

The samples were collected from Kanam, Kenya and Olduvai Gorge, Tanzania. These sites were selected because: 1) they are sites of archaeological and anthropological importance, 2) recent (Quarternary to Present) volcanic activity would have introduced many trace elements into the groundwater system in contact with the bone samples, 3) their contrasting depositional palaeoenvironments provide an opportunity to study the geochemical behaviour of multivalent elements during diagenesis of fossil bone, and 4) the alkaline soil system is conducive to good preservation of fossil bone.

Samples were studied of modern bones collected from the surface, fossil bones collected from lithified sediments, and fossil bones incorporated in calcareous concretions. Associated soil and sediment samples were also collected and, together with enclosing concretion, analysed for trace and major elements. Table 1 presents details of the samples studied.

## Sample Preparation

For instrumental neutron activation analysis (INAA), transverse serial sections of bone (and concretions enclosing bone) were prepared using a 0.1 μm diamond-impregnated wire saw. Each section was crushed prior to analysis by INAA. For electron probe microanalysis (EPMA), transverse sections of bone were mounted in araldite on 24 x 48 mm glass slides, and polished using standard lapidary techniques.

Table 1. Details of samples studied

| Sample | Site | Bed | Deposits and Environment | Taxon |
|---|---|---|---|---|
| OLD-8.10 | DK | I | Lake bottom; massive, green pyritic claystone. Bones occur within calcareous concretions. Reducing. | Reptile |
| OL-57 | DK | I | | Reptile |
| OLD-1* | TK | III | Fluvial; red hematitic sandstone. Oxidising. | Equid |
| KANAM-1+ | Kanam West | KSA | Calcareous claystone. Reducing? | Reptile |
| OLD-4.5 | FLK | Modern | Partially buried in river silt. | Antelope? |

* BMNH No. HJW-238
\+ BMNH No. HJW-241

For a description of sites at Olduvai Gorge, see Hay (1976).
For a description of Kanam West sites, see Pickford (1984).

## Analytical Techniques

INAA was used for trace element analysis of the samples using the routine analytical scheme described by Williams and Wall (in press).
Major element analyses for Ca, P, Sr and Ba in bone samples were carried out by EPMA, using a Cambridge Instrument Mark IX microprobe operated at 20 kV and $2.5 \times 10^{-8}$ A measured on a Faraday cage. Well characterised minerals and synthetic crystals were used as standards, and the raw data were fully corrected for ZAF matrix effects.

## Results and Discussion

### Trace elements Sc, La, Ce, Hf, Th, U (bone plus sediment)

Maximum and minimum concentrations for these elements in the bone cortices and associated sediments are given in Tables 2 and 3. Concentration profiles for these elements across fossil bone cortices OLD-1 and KANAM-1 are presented in Fig.1.

Fig.1 indicates that all elements, with the exception of Hf in KANAM-1 (discussed below), are enriched in the exterior portion of the bone. Comparison of these values with the low concentrations in modern bone sample OLD-4.5, Table 2, show clear evidence for post-mortal uptake of trace elements Sc, La, Ce, Th and U during diagenesis. Differences in concentration profile across fossil bone cortices were observed by Henderson et al.(1983), and the trace elements discussed here illustrate the three patterns of chemical change across fossil bone cortices as recognised by Henderson et al. (1983). The pattern types are:

(a) a trend of decreasing concentration from exterior to the interior, with a relatively small variation from maximum to minimum values,

(b) a sharp decrease in concentration from exterior to mid-cortex, with a small increase towards the central medullary cavity,

(c) an approximately U-shaped variation.

Table 2. Analytical data for fossil bone + associated sediment

| | OLD-1 | | | | KANAM-1 | | | |
|---|---|---|---|---|---|---|---|---|
| | Bone | | Sediment | | Bone | | Sediment | |
| | max. | min. | E | I | max. | min. | E | I |
| Sc[1] | 29.8 | 18.0 | 5.4 | 5.5 | 19.6 | 1.3 | 5.0 | 2.3 |
| La[1] | 412 | <.3 | 92.1 | 38.2 | 1510 | 13.3 | 246 | 107 |
| Ce[1] | 1409 | 126 | 103 | 74.0 | 2776 | 87 | 285 | 178 |
| Hf[1] | 1.14 | 0.31 | 4.7 | 3.4 | 11.5 | 3.4 | 3.0 | 6.4 |
| Th[1] | 207 | <.4 | 11.6 | 5.9 | 155 | <.5 | 22 | 22 |
| U[1] | 12.1 | 4.0 | 0.6 | 2.9 | 289 | 78 | 10.6 | 11.2 |
| Ce anom.[3] | >75 | 1.61 | 0.54 | 0.86 | 7.96 | 0.85 | 0.58 | 0.78 |
| CaO[2] | 53.2 | 45.1 | - | - | 52.1 | 47.4 | - | - |
| $P_2O_5$[2] | 36.0 | 29.1 | - | - | 35.1 | 30.9 | - | - |
| SrO[2] | 1.18 | 1.05 | 0.54* | 0.34* | 2.07 | 0.78 | 0.47* | 0.19* |
| BaO[2] | 0.35 | 0.19 | 2.81* | 3.14* | 0.24 | 0.12 | 0.13* | 0.48* |
| Sr/Ca(x1000) | 45.2 | 23.8 | - | - | 47.0 | 18.2 | - | - |

1 Concentration in ppm. Determined by INAA.
2 Concentration in weight percentage. Determined by EPMA.
3 Ce anomaly. [Ce]/[Ce*].
* Sr and Ba in sediments determined by INAA.
- not determined
E external sediment
I internal sediment

Table 3. Analytical data for fossil bone + concretions, and modern bone

| | OLD-8.10 | | | | OL-57 | | | | OLD-4.5 | | | |
|---|---|---|---|---|---|---|---|---|---|---|---|---|
| | Bone | | Concretion | | Bone | | Concretion | | Bone | | Soil | |
| | max. | min. | max. | min. | max. | min. | max. | min. | max. | min. | Es | Is |
| Sc[1] | 3.1 | 0.5 | 2.2 | 1.0 | 1.7 | 0.6 | 2.2 | 1.5 | 0.4 | <.1 | 8.1 | 10.3 |
| La[1] | 2466 | <1 | 168 | 41 | 1936 | 12 | 109 | 21 | 4.1 | <.8 | 116 | 105 |
| Ce[1] | 1852 | 22 | 252 | 74 | 1783 | 28 | 124 | 41 | 14 | <1 | 239 | 210 |
| Hf[1] | 1.3 | <.3 | 4.3 | 3.5 | 1.6 | <.3 | 4.4 | 3.1 | <.1 | <.1 | 8.6 | 9.9 |
| Th[1] | 12 | <.4 | 14.6 | 8.4 | 27 | <.5 | 16.5 | 8.1 | 0.4 | <.4 | 16.4 | 13.9 |
| U[1] | 349 | 174 | 7.4 | 1.9 | 388 | 180 | 18.0 | 4.0 | 11.4 | 0.3 | 3.7 | 3.9 |
| Ce anom.[3] | 2.31 | 0.33 | 2.18 | 0.43 | 3.4 | 0.47 | 1.07 | 0.57 | - | - | 0.97 | 0.91 |
| CaO[2] | 51.0 | 47.2 | - | - | 51.3 | 45.2 | - | - | 47.4 | 43.7 | - | - |
| $P_2O_5$[2] | 34.0 | 31.8 | - | - | 33.2 | 30.4 | - | - | 35.1 | 31.3 | - | - |
| SrO[2] | 0.87 | 0.71 | 0.22* | 0.07* | 0.82 | 0.65 | 0.22* | 0.06* | 0.84 | 0.05 | 0.17* | 0.10* |
| BaO[2] | 0.25 | 0.17 | 0.36* | 0.06* | 0.30 | 0.24 | 0.06* | 0.03* | 0.10 | <.02 | 0.18* | 0.12* |
| Sr/Ca(x1000) | 22.0 | 16.5 | - | - | 20:5 | 15.8 | - | - | 22.4 | 0.7 | - | - |

Legend as for Table 2 except:-

Es external soil
Is internal soil

The overall pattern types for the elements Sc, La, Ce, Th and U are similar in bone samples from contrasting depositional palaeoenvironments, although absolute concentrations may differ (see Williams and Marlow 1987, for a discussion on U and Th, and below). In those elements discussed here, Sc, Ce and U conform to type (a), La and Th to type (b).

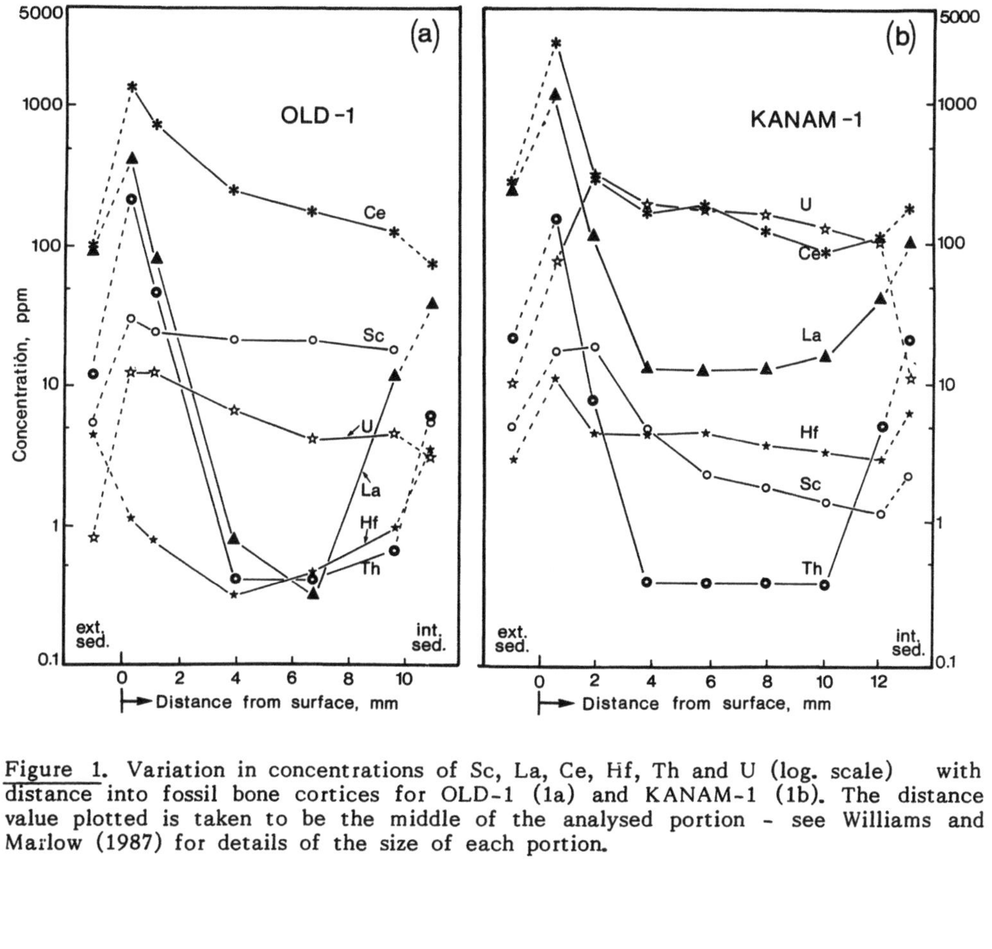


Figure 1. Variation in concentrations of Sc, La, Ce, Hf, Th and U (log. scale) with distance into fossil bone cortices for OLD-1 (1a) and KANAM-1 (1b). The distance value plotted is taken to be the middle of the analysed portion - see Williams and Marlow (1987) for details of the size of each portion.

For Hf however, the pattern type changes from (c) in OLD-1 (oxidising palaeoenvironment) to (a) in KANAM-1 (reducing plaeoenvironment). In addition, the absolute concentration of Hf is an order of magnitude higher in KANAM-1 than in OLD-1, and is greater than in the external sediment, Table 2. The differences in Hf behaviour between the Kanam and Olduvai Gorge sites is likely to have resulted from different concentration levels, and in the mobility of this element in the groundwater at the two sites. The geochemistry of Hf is similar to that of Zr (e.g. Mason and Moore,

1982) which is known to be present in solution at unusually high concentrations in hot springs in the Kanam area (Henderson et al. 1987). Hot springs would contribute to, and enrich the concentration of Zr ( and by analogy Hf) in the local groundwater system at Kanam, and these elements would be in a form locally available for increased uptake by fossil bone during diagenesis.

The Sc concentration in sample OLD-1 is approximately one order of magnitude higher than in KANAM-1, and is the result of Sc entering Fe-Mn minerals which have precipitated in pore spaces in the sample (OLD-1) from the oxidising environment (Williams and Potts, in press). Fe-Mn minerals are absent from the KANAM-1 sample as the redox potential of the groundwater system associated with this sample was sufficiently reducing to maintain Fe and Mn as 2+ ions in solution. In contrast, U concentrations in sample OLD-1 are lower than in KANAM-1. As a multivalent element, U is sensitive to changes in the redox potential of the groundwater, and the order of magnitude difference in U concentration between the two samples can also be attributed to redox potential differences. The oxidised species ($UO_2^{2+}$) is more soluble than the reduced species ($U^{3+}$) in an aqueous medium, and Williams and Marlow (1987) argued that fossil bones from an oxidising palaeoenvironment contain lower U concentrations than bones from a reducing palaeoenvironment.

The multivalent rare earth element (REE) Ce has two oxidation states, 3+ and 4+. However, the adjacent elements La and Pr have only one oxidation state (3+), and the extent to which Ce occurs in the oxidised (4+) state can be assessed by comparing the chondrite-normalised values of Ce with La and Pr. A Ce anomaly value, Ce / Ce* can be calculated, where Ce is the Ce concentration, and Ce* is the value obtained at the Ce plotted position by linear interpolation between La and Pr - see for example Henderson (1984) for a description of normalisation procedures, and a critique of the Ce anomaly. A Ce / Ce* value greater than 1 (positive anomaly) indicates an excess of Ce relative to adjacent REE, and a value less than 1 (negative anomaly) indicates a depletion of Ce. The Ce anomalies are positive in both fossil bone cortices, but the magnitude of this anomaly is different from bones from other palaeoenvironments. In OLD-1 from an oxidising environment, the maximum Ce anomaly value is significantly larger ( >75) than in KANAM-1 (7.96), from a reducing environment (Table 2). Differences in Ce anomaly have also been observed for different beds in Olduvai Gorge (Williams and Henderson, in preparation), and reflect also the redox potential of the palaeoenvironment. The observed differences in Ce anomaly are due to faster rates of uptake and equilibrium of $Ce^{4+}$ relative to the trivalent REE (Henderson et al. 1983), and the larger Ce anomaly values in the more oxidised palaeoenvironments result from a higher proportion of $Ce^{4+}$ to $Ce^{3+}$ ions in the groundwater.

Bone plus concretion

Calcareous concretions,some containing fossil bones, are common in the lacustrine deposits of Bed I at Olduvai Gorge (Hay 1976). Such concretions usually form early in the diagenetic process as a result of precipitation from the pore/groundwater in the lake sediment (Tucker 1981). Since trace elements from the groundwater system have been incorporated into fossil bone during diagenesis, it is instructive to look at the distribution of those elements in the concretion precipitated from the same groundwater system and (presumably) during the same diagenetic period. Fossiliferous concretions OL-57 and OLD-8.10 were sectioned and analysed, the results are summarised in Table 2, and the distribution of elements Sc, La, Ce, Hf and Th for OLD-8.10 are shown in Fig.2. Fig.2 shows a symmetrical distribution of Sc, La, Ce and Th in the two half-cortices of the bone, with similar patterns to those from OLD-1 and KANAM-1, and with the highest concentrations of these elements closest to the external bone surface. Like sample OLD-1, the concentration of Hf in OLD-8.10 bone sections is lower than that in the adjacent concretion, indicating very little uptake of this element into the bone during diagenesis.

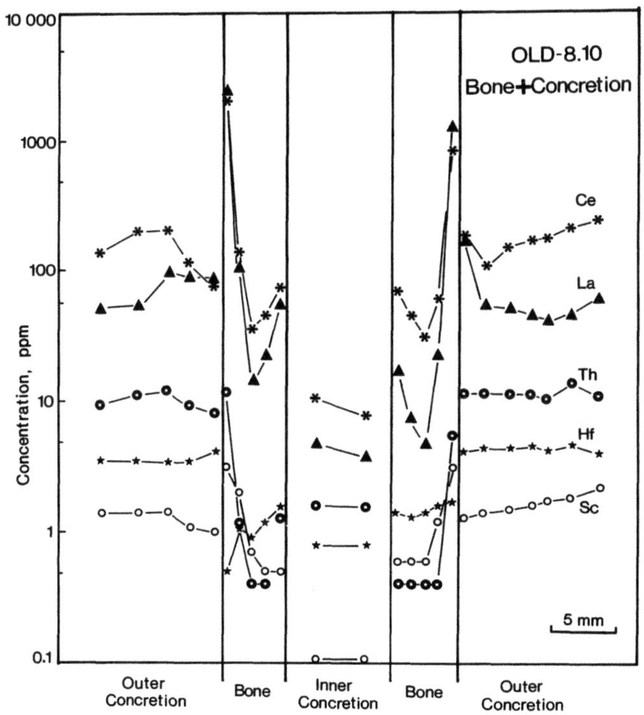


Figure 2. Variation in concentration of Sc, La, Ce, Hf and Th (log.scale) with distance across fossil-bearing concretion OLD-8.10.

In the concretion enclosing the bone, Hf and Th concentrations show little or no change with distance away from the bone (Fig. 2), suggesting that these elements were not present as soluble species in the groundwater system during diagenesis. It is likely that Hf forms pseudo-colloidal species similar to those for Th (e.g. Airey and Ivanovich, 1986), or are adsorbed onto clay minerals, and are not directly available for uptake into fossil bone, although Th can be indirectly incorporated into fossil bone (Williams and Marlow 1987).

Sc shows a slight, but significant, systematic decrease in concentration in the concretion towards the bone, with a corresponding enrichment in the bone's outer surface. It is suggested, that Sc ions were present in the groundwater system (Fig.2). Sc was incorporated into the bone during diagenesis, causing its local depletion from the surrounding groundwater. Subsequent precipitation from the groundwater to form the concretion resulted in a zone adjacent to the bone depleted in Sc.

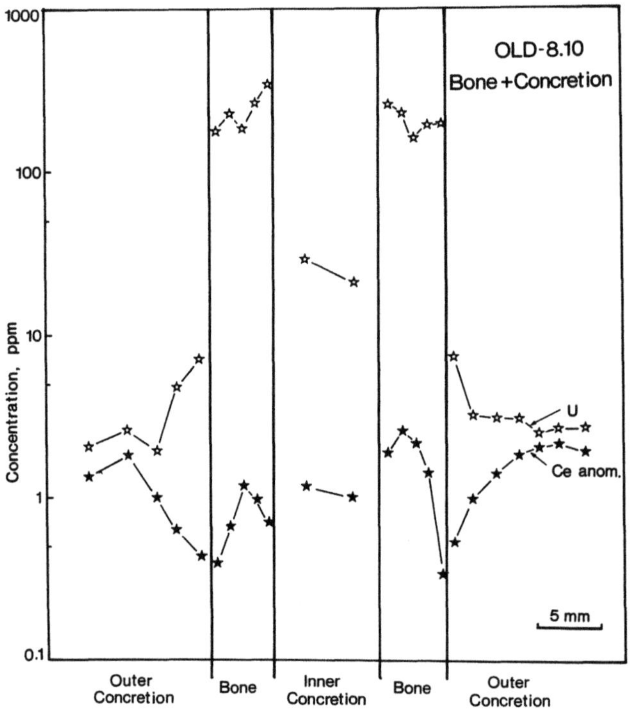


Figure 3. Variation in concentration of U and Ce anomaly ( Ce / Ce* ) (log.scale) with distance across fossil-bearing concretion OLD-8.10

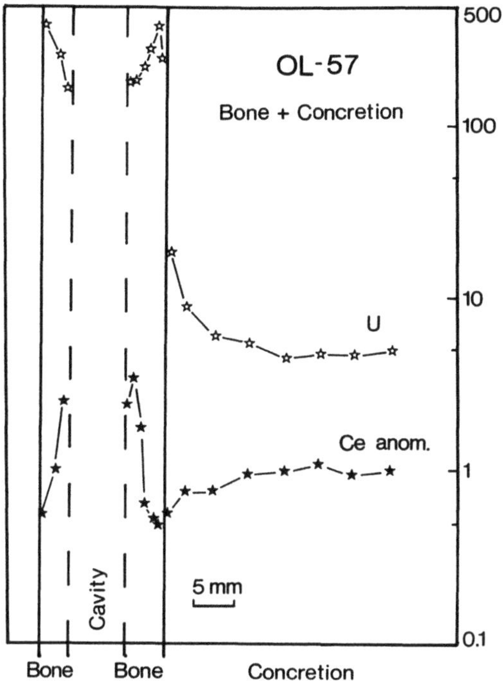


Figure 4. Variation in concentration of U and Ce anomaly ( Ce / Ce* ) (log.scale) with distance across fossil-bearing concretion OL-57

Although the distribution profiles for La and Ce show systematic variations in the bone cortices (Fig.2), there is no corresponding systematic variation in the adjacent concretion. However, changes in the La to Ce ratio, i.e. a function of the Ce anomaly are present in the concretion, and reflect the fact that Ce, unlike La, is a multivalent element.

The multivalent elements U and Ce (represented by the Ce anomaly, Ce / Ce* ) have significant variations in both the bones and enclosing calcareous concretions in samples OLD-8.10 (Fig. 3) and OL-57 (Fig. 4). In concretions from both samples, the U concentration increases, and the Ce anomaly decreases towards the bone surfaces, with the highest U concentrations and lowest (negative) Ce anomalies closest to the bone surfaces. It is suggested that these systematic variations in U and Ce anomaly are a direct result of a corresponding change in the redox potential of the groundwater immediately adjacent to the fossilising bone. The breakdown of organic components during early stages of fossilisation is likely to produce locally a reducing micro-environment (e.g. Rae and Ivanovich 1986) which, in modern bones can accumulate high concentrations of U (Williams and Marlow 1987). It is likely that the decaying bone produced a zone of reduced redox potential in the groundwater, which has affected the uptake of multivalent elements into the calcareous concretion precipitated from this groundwater. Thus, U concentrations are highest, and Ce anomalies lowest in the most reducing parts of the bone-groundwater system, i.e. at their interface. Figs. 3 and 4 show that this zone of reduced redox potential in the groundwater must have extended several mms from the bone.

## Sr and Ba

There have been many studies which attempt to use the distribution of trace elements in fossil bone to determine palaeonutritional information of past populations. Although a variety of elements has been examined, only a few have been accepted as reliable indicators of pre-mortem conditions, e.g. Sr, Ba, Zn, Mn (Price et al. 1985). The majority of palaeonutritional studies have however concentrated on Sr (and to a lesser extent, Ba) and its application to palaeodietary research (see Sillen and Kavanagh, 1982, for a review). Although diagenetic processes have been recognised as potentially obscuring the in vivo trace element composition in bone, some correction procedures have been suggested to identify (e.g. Lambert et al. 1985), and to minimise diagenetic effects (e.g. Schoeninger 1982; Patterson et al. 1987). One such procedure for Sr is to use the Sr/Ca ratio. Since Sr substitutes for Ca in the bone apatite crystralstructure, any dissolution of bone will affect Sr and Ca equally and the Sr/Ca ratio will remain unchanged (Schoeninger 1982). Some recent studies (Nelson et al., 1986; Klepinger et al. 1986) have indicated that diagenesis, including contamination by microorganisms (Grupe and Piepenbrink, this volume), may severely affect the distribution of several elements, including Sr. Consequently, Sr may not be useful an indicator of palaeonutrition as has been previously suggested.

To asses the applicability of Sr and Ba as potentially useful palaeonutritional indicator elements in the E.African sites of Olduvai Gorge, Tanzania and Kanam,Kenya, these elements (together with Ca and P) were determined by EPMA. Maximum and minimum concentrations of Ca, P, Sr and Ba in the bone cortices are given in Tables 2 and 3, and plots of Sr/Ca(x1000) against distance acrosss fossil bone cortices are presented in Fig. 5, and across modern bone cortex (OLD-4.5) in Fig. 6.

The distribution profiles of Sr/Ca(x1000) in the fossil bones, Fig. 5, show variable, but relatively flat profiles across cortices. Of note are the exceptionally high Sr/Ca (x1000) values (up to 47.0) that result from high Sr concentration in these bones (up to 2.07% SrO). It was initially thought that such high concentration levels were in vivo concentrations due to intake of Sr-rich dietary material. However, examination of the Sr/Ca(x1000) profile of a modern bone (OLD-4.5), collected from the surface

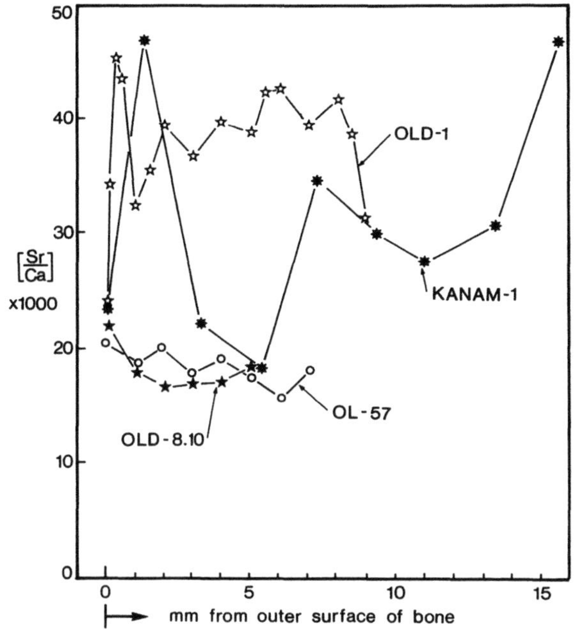


Figure 5. Variation of Sr/Ca(x1000) with distance into fossil bone cortices for OLD-1, KANAM-1,OLD-8.5 and OL-57

at Olduvai Gorge, Fig. 6, shows clear and unambiguous evidence of extreme and rapid post-mortem uptake of Sr. The high nitrogen content of OLD-4.5 (4-5% N,C.A.Marlow unpublished data), and well preserved features (OLD-4.5 is an intact limb bone), indicate that this bone has been on the surface for a very short time, probably less than one year. The bone displays a surface colouration, from dark brown where totally covered by soil, through light brown (partly covered) to white (uncovered).

Distribution profiles for Sr/Ca(x1000) were obtained across the cortex at positions corresponding to the colour variations, Fig. 6. In each of the three profiles, the distribution of Sr/Ca(x1000) is asymmetrical across the cortex.

Highest concentrations of Sr (and Sr/Ca(x1000)) occur at the outer surface layers (up to 0.22% SrO). Sr (and Sr/Ca(x1000)) values then decrease rapidly over distance of less than 0.5 mm to 0.05% SrO, and increase again on approaching the medullary cavity. Ca concentrations are relatively constant across the whole cortex. It is noteworthy that in the outer surface layers, the Sr concentrations correlate with the bone colouration (Fig. 6); the darkest (soil covered) region contains the highest Sr (and

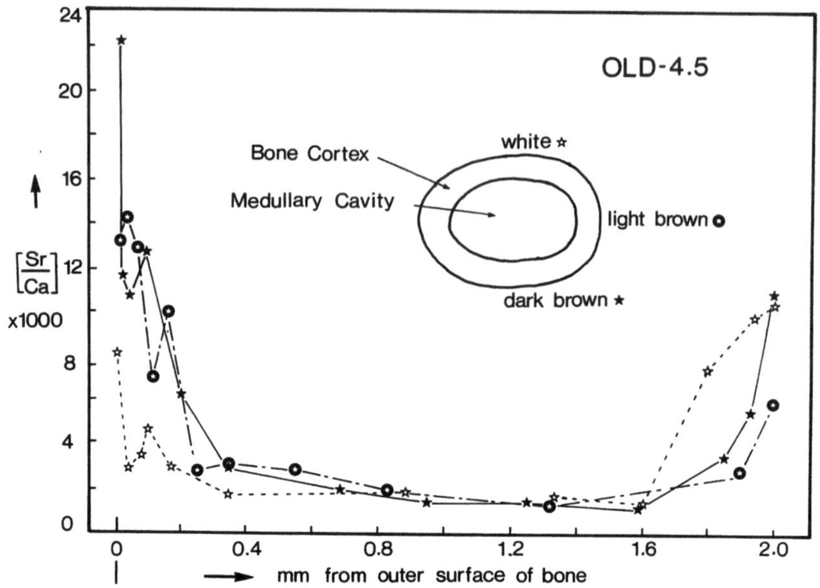


Figure 6. Variation of Sr/Ca(x1000) with distance into the modern bone cortex, OLD-4.5. White is uncovered; light brown, partly covered; dark brown, totally covered by soil.

Sr/Ca(x1000)) values, and the lightest (uncovered by soil) the lowest values. This overall pattern and correlation with colour is very similar to the U distribution profile for the same bone section (Williams and Marlow 1987), and illustrates the considerable and rapid post-mortem uptake of both Sr and U. The Sr content of the soil associated with the bone is significantly lower than in the outer surface layers of the bone (Table 3), suggesting that $Sr^{2+}$ ions were present in the groundwater and available for incorporation into bone during early diagenesis.

The data for Sr clearly show that this element cannot be used as an indicator element for palaeodietary studies in the important palaeoanthropological sites of Olduvai Gorge, Tanzania and Kanam, Kenya. These results are consistent with those from Nelson et al.(1986) for other sites in America, who state "Our results suggest that postmortem alteration of dietary tracers in the inorganic phases of bone may be a problem at all archaeological sites and must be evaluated in each case."

In contrast to Sr, Ba appears to be less affected by early diagenesis, with concentrations in the modern bone sample OLD-4.5 ranging from 0.10% to 0.02% BaO. The distribution profiles for Ba in the fossil bones (not plotted) are similar to those for Sr, i.e. relatively flat, but the concentrations of Ba are both much lower than

Sr, and also have a smaller range for the different samples, Tabels 2 and 3. The flat distribution profiles for Sr and Ba across the fossil bones cortices indicate that both these elements have achieved equilibrium between the bone and the groundwater; however, Ba is incorporated into bone at a lower rate than Sr.

## Conclusions

1) All elements studied here show evidence of post-mortal uptake into bone, but to varying degrees. 2) The major factors predominating in elemental uptake have been identified as element availability in the groundwater, and the redox potential of the palaeoenvironment. 3) The distribution of multivalent elements Ce and U in both fossil bone and concretions enclosing fossil bone provide information on the palaeoenvironment. 4) Sr, like U, is rapidly incorporated into bone during very early stages of diagenesis, and cannot be used as an indicator element for palaeonutritional studies. 5) Ba and Hf are least affected by diagenesis, although both these elements are incorporated to some extent into fossil bone.

## Acknowledgements

This paper summarises some of the findings of a study on element mobility and uptake during diagenetic alteration of fossil bones from Olduvai Gorge, Tanzania and Kanam, Kenya. The project was a team effort involving C A Marlow (Dept.of Anthropology, University of Durham), P Henderson and T I Molleson (British Museum, Natural History) to whom I am indebted for the use of some unpublished data and numerous discussions.

## References

Airey PL, Ivanovich M (1986) Chemical analogues of high-level radioactive waste repositories. Chem Geol 55: 203-213

Grupe G, Piepenbrink H (1987) Trace element contaminations in excavated bones by microorganisms. In: Grupe G, Herrmann B (eds) Trace Elements in Environmental History, Springer Berlin Heidelberg New York, pp 103-112

Henderson P (1984) Geochemical properties and abundances of the rare earth elements. In: Henderson P (ed) Rare Earth Element Geochemistry, Elsevier Amsterdam Oxford New York, pp 1-32

Henderson P, Marlow CA, Molleson TI, Williams CT (1983) Patterns of chemical change during bone fossilization and their significance. Nature 306: 358-360

Henderson P, Pickford M, Williams CT (1987) A geochemical study of rocks and spring waters at Kanam and Kanjera, Kenya, and the implications concerning element mobility and uptake. J Afr Earth Sci 6: 221-227

Klepinger LL (1984) Nutritional assessment from bone. Ann Rev Anthrop 13: 75-96

Klepinger LL, Kuhn JK, Williams Wendell S (1986) An elemental analysis of archaeological bone from Sicily as a test of predictability of diagenetic change. Am J Phys Anthrop 70: 325-331.

Lambert JB, Simpson SV, Szpunar CB, Buikstra JE (1985) Bone diagenesis and dietary analysis. J Hum Evol 14: 477-482.

Mason B, Moore CB(1982) Principles of Geochemistry (4th edition). John Wiley and Sons New York Chichester, pp 344ff

Nelson BK, DeNiro MJ, Schoeninger MJ, Paolo DJ, Hare PE (1986) Effects of diagenesis on strontium, carbon, nitrogen and oxygen concentration and isotopic composition of bone. Geochim Cosmochim Acta 50: 1941-1949.

Patterson CC, Shirahata H, Ericson JE (1987) Lead in ancient human bones and its relevance to historical developments of social problems with lead. Sci Tot Environ 61: 167-200

Pickford M (1984) Kenya Palaeontology Gazetter, Vol 1, Western Kenya. Spec Publ Nat Mus Kenya 1984: 1-282

Price TD, Schoeniger MJ, Armelagos GJ (1985) Bone chemistry and past behaviour: an overview. J Hum Evol 14: 419-447

Rae AM, Ivanovich M (1986) Successful application of uranium series dating of fossil bone. Appl Geochem 1: 419-426

Schoeninger MJ (1982) Diet and evolution of modern human form in the Middle East. Am J Phys Anthrop 58: 37-57

Sillen A, Kavanagh M (1982) Strontium and palaeodietary research: a review. Yearb Phys Anthrop 25: 67-90

Tucker ME (1981) Sedimentary Petrology: an Introduction. Geoscience Texts, Vol 3, Blackwell Scientific Publ Oxford London, pp 252ff

Williams CT, Marlow CA (1987) Uranium and thorium distributions in fossil bones from Olduvai Gorge, Tanzania and Kanam, Kenya. J Arch Sci 14: 287-309

Williams CT, Potts PJ (in press) Element distribution maps in fossil bones by fission track analysis and neutron activated-induced beta autoradiography. Archaeometry

Williams CT, Wall F (in press) An INAA scheme for the routine determination of 27 elements in geological and archaeological samples. In: Hughes MJ, Cowell M, Hook DR (eds) Neutron Activation Analysis and Plasma-arc Spectrometry (ICP) in Archaeology. Brit Arch Rep

# MULTIELEMENTARY ANALYSIS OF TRACE ELEMENTS AND PRELIMINARY RESULTS ON STABLE ISOTOPES IN TWO ITALIAN PREHISTORIC SITES. METHODOLOGICAL ASPECTS.

Paolo Francalacci and Silvana Borgognini Tarli
Istituto di Antropologia e Paleontologia Umana
University of Pisa
Via S. Maria, 55
56100 Pisa, Italy

## Introduction

This research, born of the interest of paleobiochemical investigation and its potential value in studying the way of life of past human populations, aimed at overcoming some of the main difficulties so far encountered in the application of trace element dosage to dietary reconstruction. These are:

- the possible existence of intra-individual variations (due to bone and/or kind of tissue, compact or cancellous);
- the need to evaluate the effect of diagenetic alterations, taking into account the possibility of sample-specific mechanisms, and to set up standardized control procedures;
- the choice of the more suitable elements, considering that a multielementary approach can be more informative than the dosage of a single element;
- the possibility to exploit the availability of two independent techniques applicable to paleonutritional reconstruction (trace element and stable isotope dosage) for a reciprocal control and support of the results.

A large sample of complete and less valuable skeletons would make for a more simple approach to the above problems, nonetheless we preferred to use human and animal material in a typical situation, that is from archaeological sites with variously preserved and anthropologically interesting human remains and with unevenly distributed faunal remains.

## Materials and methods

The sample comes from the Epigravettian and Middle Neolithic levels in Arene Candiede cave (Western Liguria, Italy) and from Mesolithic and Neolithic levels in the Uzzo cave (Western Sicily, Italy). The samples from Arene Candide cave date to about 11,000 BP (Bietti 1987) and to 5,500 - 6,000 BP (Maggi 1977, Tiné 1975) respectively. The samples from Uzzo cave date to 9-10,000 BP and to 6,750 ±70BP (Meulengracht et al. 1981, Piperno 1985) respectively.
The sample composition is shown in table 1.

| | Arene Candide | | Uzzo | |
|---|---|---|---|---|
| | Epigravettian | Neolithic | Mesolithic | Neolithic |
| Herbivores | 15 | 28 | 15 | 1 |
| Sus | - | 5 | 8 | 2 |
| Omnivores | 3 | 13 | - | - |
| Carnivores | 1 | 4 | 9 | 3 |
| Humans | 10(8) | 11(7) | 6(5) | 1(1) |
| Total | 29 | 66 | 38 | 7 |

Table 1. Composition of the sample. The faunal subsamples include the following genera: Herbivores - Cervus, Bos, Ovis, Capra, Rupicapra, Capreolus; Omnivores - Meles, Ursus; Carnivores - Felis, Martes, Vulpes.
Note: The "humans" line reports the number of bones analysed and, in brackets, the number of individuals.

As the material derives from excavations completed some time ago, it was not possible to perform analyses of the soil in proximity of a bone in primary deposition. A recent excavation of a burial from Arene Candide allowed this control in one case only, in which four soil samples were taken at 0, 5, 10 and 20 cm from a human occipital bone.
A variant of the method proposed by Szpunar et al. (1978) was followed for the preparation of the sample: The drying of the acidic hydrolysate of the bone was excluded in order to avoid the precipitation of the silicates and consequently the filtration phase, a possible source of pollution.
The samples were analysed using a Perkin Elmer ICP 6 000 Spectrometer. The elements measured were: Ba, Cu, Mg, Mn, Sr, V and Zn.
Inductively Coupled Plasma Atomic Emission Spectrometry was preferred to the more commonly used Atomic Absorption Spectrometry because of its lower susceptibility to interferences (Brown & Blakely 1985, Price 1985, Price et al. 1985). The instrumental error range, calculated with standard solutions, was ±10%.
A subsample of human and herbivorous remains from Arene Candide cave was analysed for the $^{13}C/^{12}C$ and $^{15}N/^{14}N$ ratio. After collagen extraction and combustion, the samples were analysed using a Mass Spectrometer VG SIRA 9.

Results and Discussion

Table 2 shows the results of tests on different bones of the same individual.

| Indiv. | Bone | Ba | Cu | Mg | Mn | Sr | V | Zn |
|---|---|---|---|---|---|---|---|---|
| AC 5 | vert./rib | 0.59 | 1.06 | 1.11 | 1.08 | 0.99 | 0.97 | 1,01 |
| AC 10 | skull/rib | 0.18 | 0.74 | 0.60 | 1.89 | 0.65 | 0.57 | 0.60 |
| AC 10(*) | skull/rib | 0.15 | 0.68 | 0.63 | 1.71 | 0.66 | 0.43 | 0.64 |
| AC T1 | vert./rib | 0.94 | 0.86 | 1.02 | 3.07 | 0.92 | 1.13 | 0.84 |
| AC X | vert./rib | 0.55 | 0.56 | 0.85 | 0.25 | 0.77 | 0.77 | 0.73 |
| AC T2 | hip b./rib. | 0.56 | 0.40 | 1.78 | 0.18 | 0.51 | 0.31 | missing |
| AC T2 | sacr./rib. | 0.85 | 0.87 | 1.26 | 0.99 | 0.73 | 0.80 | missing |
| AC T2 | hip b./sacr. | 0.66 | 0.45 | 1.40 | 0.19 | 0.69 | 0.39 | 0.82 |
| U 7 | hand/rib | 0.77 | 0.71 | 0.57 | 0.51 | 0.79 | 0.36 | 0.54 |

Table 2. Ratios of amounts of trace elements dosed in different bones of the same individual. (*): Repeated test on a new preparation from the same bone powders.

Only AC 5 and AC T1 gave similar results, while other individuals showed different amounts of trace elements. Although some regularity can be detected, the differences are not predictable: Preferentially higher amounts are observable in ribs as compared to other bones, for example , but not in every case.

It should be noted that the differences observed are not due to experimental errors, because repeated tests on the same bone (performed in 16 cases, of which 4 are included in the subsample of table 2) gave concordant results, within the range of the instrumental error.

These results are rather similar to those obtained by Klepinger et al. (1986), who found a remarkable und unpredictable intraindividual variation. A high intraindividual variability was also noted by Herrmann and Grupe (1987).

An explanation could lie in the dishomogeneity of the bone tissue (Klepinger et al. 1986) or in metabolic differences in the various body districts, but this would not account for a different behaviour from the same skeletal element (such as ribs and vertebrae of this sample). Diagenetic changes due to various species of microorganisms (Grupe and Piepenbrink 1987) and/or to localized differences in the geochemical environment of the burial, could increase the intraindividual variability.

Tables 3-5 show the results of univariate statistical analysis. In order to overcome problems related to the normality of distributions, both parametric and non parametric tests were applied.

ARENE CANDIDE - EPIGRAVETTIAN

| El. | Subs. | n° | $\bar{x}$(ppm) | c.v. | P≤.01 | .05 ≥ P >.01 | N.T. |
|---|---|---|---|---|---|---|---|
| | Herb | 15 | 173.0 | 31.5 | - | Herb-Omn (t F U) | - |
| Ba | Omn | 3 | 99.0 | 26.0 | | | |
| | Carn | 1 | 69.9 | - | | | |
| | Hum | 10 | 155.2 | 56.0 | | | |
| | Herb | 15 | 14.66 | 31.1 | - | - | Herb-Omn |
| Cu | Omn | 3 | 21.25 | 53.1 | | | |
| | Carn | 1 | 15.68 | - | | | |
| | Hum | 10 | 16.21 | 33.7 | | | |
| | Herb | 15 | 2684.1 | 9.8 | Herb-Omn (t F) | Herb-Omn (u) | - |
| Mg | Omn | 3 | 2088.9 | 15.4 | Herb-Hum (t F U H) | | |
| | Carn | 1 | 2013.9 | - | | | |
| | Hum | 10 | 1967.9 | 20.8 | | | |
| | Herb | 15 | 70.4 | 130.8 | - | - | Omn-Hum |
| Mn | Omn | 3 | 53.7 | 61.3 | | | |
| | Carn | 1 | 231.7 | - | | | |
| | Hum | 10 | 152.2 | 99.7 | | | |
| | Herb | 15 | 229.7 | 11.5 | Herb-Hum (t F U H) | - | Herb-Omn |
| Sr | Omn | 3 | 195.2 | 34.7 | | | Omn-Hum |
| | Carn | 1 | 174.8 | - | | | |
| | Hum | 10 | 156.6 | 18.6 | | | |
| | Herb | 15 | 7.44 | 31.6 | - | Herb-Hum(t U H) | Herb-Omn |
| V | Omn | 3 | 6.60 | 2.9 | | | Omn-Hum |
| | Carn | 1 | 4.94 | - | | | |
| | Hum | 10 | 5.56 | 36.0 | | | |
| Zn | Herb | 15 | 138.4 | 19.7 | Herb-Omn (t F) | Herb-Omn (U H) | Herb-Hum |
| | Omn | 3 | 200.9 | 22.2 | Herb-Hum (U H) | | |
| | Carn | 1 | 223.3 | - | | | |
| | Hum | 9 | 276.6 | 37.0 | | | |

Table 3. Arene Candide - Epigravettian: Univariate statistics.
Legend: t = Student's t; F = Analysis of variance; U = Mann-Whitney's test; H = Kruskall-Wallis' test; N.T. = Cases non testable with parametric tests because the condition of variance homogeneity was not respected.

ARENE CANDIDE - MIDDLE NEOLITHIC

| El. | Subs. | n° | $\bar{x}$ (ppm) | c.v. | $P \leq .01$ | $.05 \geq P > .01$ | N.T. |
|---|---|---|---|---|---|---|---|
| | Herb | 27 | 246.5 | 27.8 | Herb-Carn (t U H) | Herb-Carn (F) | Omn-Hum |
| | Sus | 5 | 238.5 | 13.2 | Herb-Hum (t F U H) | Sus-Carn (F U) | |
| Ba | Omn | 13 | 218.9 | 33.2 | Sus-Carn (t H) | Carn-Hum (U H) | |
| | Carn | 4 | 158.3 | 25.3 | Sus-Hum (t F U H) | | |
| | Hum | 11 | 84.9 | 47.5 | Omn-Hum (U H) | | |
| | | | | | Carn-Hum (t F) | | |
| | Herb | 28 | 12.57 | 48.3 | Herb-Carn (t F H) | Herb-Carn (U) | Herb-Omn |
| | Sus | 5 | 11.96 | 38.6 | Herb-Hum (U) | Herb-Hum (H) | Herb-Hum |
| Cu | Omn | 13 | 14.72 | 69.1 | Sus-Carn (t F U H) | | |
| | Carn | 4 | 26.52 | 24.5 | Omn-Carn (T U H) | | |
| | Hum | 11 | 19.54 | 50.8 | | | |
| | Herb | 28 | 2345.9 | 26.1 | Omn-Hum (U H) | Herb-Hum (t F U H) | |
| | Sus | 5 | 2456.7 | 13.9 | | Sus-Carn (t F H) | |
| Mg | Omn | 13 | 2345.7 | 18.7 | | Sus-Hum (t F U H) | |
| | Carn | 4 | 1903.8 | 12.0 | | Omn-Carn (t) | |
| | Hum | 11 | 1914.1 | 24.7 | | Omn-Hum (t F) | |
| | Herb | 28 | 64.1 | 72.4 | Herb-Sus (t F U H) | Herb-Hum (U H) | Herb-Omn |
| | Sus | 5 | 151.5 | 36.8 | | Sus-Carn (U H) | Herb-Carn |
| Mn | Omn | 13 | 98.1 | 90.8 | | | Herb-Hum |
| | Carn | 4 | 53.2 | 21.0 | | | Sus-Carn |
| | Hum | 11 | 110.6 | 63.6 | | | Omn-Carn |
| | | | | | | | Carn-Hum |
| | Herb | 28 | 210.5 | 17.1 | Herb-Hum (t F U H) | Herb-Sus (t U H) | Herb-Omn |
| | Sus | 5 | 240.3 | 12.4 | Sus-Hum (t F U H) | Herb-Omn (U H) | Omn-Hum |
| Sr | Omn | 13 | 242.1 | 28.8 | Omn-Hum (U H) | Carn-Hum (U) | |
| | Carn | 4 | 199.9 | 17.3 | Carn-Hum (t F H) | | |
| | Hum | 11 | 136.1 | 23.0 | | | |
| | Herb | 28 | 7.25 | 46.2 | - | Herb-Hum (t U) | Sus-Omn |
| | Sus | 5 | 6.49 | 18.1 | | Sus-Carn (t) | |
| V | Omn | 13 | 5.96 | 52.5 | | | |
| | Carn | 4 | 4.58 | 36.2 | | | |
| | Hum | 11 | 5.13 | 44.4 | | | |
| | Herb | 18 | 115.4 | 26.4 | Herb-Omn (t F U H) | Carn-Hum (U H) | Herb-Sus |
| | Sus | 5 | 132.1 | 3.4 | Herb-Carn (t F U H) | | Herb-Hum |
| Zn | Omn | 13 | 152.2 | 18.8 | Herb-Omn (U H) | | Sus-Omn |
| | Carn | 4 | 205.1 | 9.8 | Sus-Carn (U H) | | Sus-Carn |
| | Hum | 10 | 335.5 | 32.6 | Sus-Hum (U H) | | Sus-Hum |
| | | | | | Omn-Carn (U H) | | Omn-Hum |
| | | | | | Omn-Hum (U H) | | Carn-Hum |

Table 4. Arene Candide - Middle Neolithic: univariate statistics.
Legend as in Tab. 3.

UZZO - MESOLITHIC

| El. | Subs. | n° | $\bar{x}$(ppm) | c.v. | P≤.01 | .05 ≥ P>.01 | N.T. |
|---|---|---|---|---|---|---|---|
| | Herb | 15 | 136.2 | 21.4 | Herb-Carn (t F U H) | Herb-Sus (t F U H) | - |
| Ba | Sus | 8 | 105.0 | 27.4 | Herb-Hum (t F U H) | Sus-Hum (F U H) | |
| | Carn | 9 | 87.9 | 19.4 | Sus-Hum (t) | | |
| | Hum | 7 | 67.9 | 36.4 | | | |
| | Herb | 14 | 12.82 | 41.1 | - | Herb-Carn (t) | - |
| Cu | Sus | 8 | 13.41 | 20.1 | | Sus-Carn (t) | |
| | Carn | 9 | 17.36 | 32.9 | | | |
| | Hum | 7 | 14.33 | 31.7 | | | |
| | Herb | 15 | 2723.9 | 19.1 | Sus-Hum (t) | Herb-Sus (t F) | Sus-Carn |
| Mg | Sus | 8 | 3321.2 | 26.0 | | Herb-Hum (t) | |
| | Carn | 9 | 2818.9 | 23.7 | | Sus-Hum (F U H) | |
| | Hum | 7 | 2190.1 | 33.9 | | Carn-Hum (t) | |
| | Herb | 15 | 53.4 | 84.6 | Sus-Hum (t) | Herb-Hum (t F H) | Sus-Carn |
| Mn | Sus | 8 | 38.9 | 81.3 | | Sus-Hum (F U H) | |
| | Carn | 9 | 90.4 | 78.6 | | | |
| | Hum | 7 | 125.8 | 60.2 | | | |
| | Herb | 15 | 244.9 | 25.8 | - | Herb-Sus (t F U H) | - |
| Sr | Sus | 8 | 191.4 | 17.4 | | Herb-Carn (t F U H) | |
| | Carn | 9 | 191.5 | 16.8 | | Herb-Hum (t) | |
| | Hum | 7 | 192.1 | 26.2 | | | |
| | Herb | 15 | 8.68 | 29.6 | - | - | Herb-Sus |
| V | Sus | 8 | 9.18 | 57.6 | | | Herb-Hum |
| | Carn | 9 | 8.14 | 31.1 | | | Sus-Carn |
| | Hum | 7 | 11.99 | 46.1 | | | Carn-Hum |
| | Herb | 15 | 120.7 | 32.7 | Herb-Carn (t F U) | Herb-Carn (H) | Herb-Sus |
| Zn | Sus | 8 | 116.5 | 16.6 | Herb-Hum (t F U H) | Sus-Carn (U H) | Sus-Carn |
| | Carn | 9 | 172.1 | 27.1 | Sus-Hum (U H) | | Sus-Hum |
| | Hum | 6 | 191.9 | 21.9 | | | |

UZZO - NEOLITHIC

| Subs. | n° | Ba | Cu | Mg | Mn | Sr | V | Zn |
|---|---|---|---|---|---|---|---|---|
| Herb | 1 | 131.1 | 17.1 | 3888.3 | 48.3 | 210.4 | 25.1 | 62.8 |
| Sus | 2 ($\bar{x}$) | 138.9 | 18.6 | 3026.8 | 152.8 | 202.1 | 25.1 | 95.1 |
| Carn | 3 ($\bar{x}$) | 91.0 | 16.3 | 2923.2 | 68.7 | 178.9 | 21.0 | 163.0 |
| Hum | 1 | 62.5 | 13.9 | 2771.5 | 83.4 | 214.0 | 17.0 | 75.5 |

Table 5. Uzzo - Mesolithic and Neolithic : univariate statistics.
Legend as in Tab. 3.

The faunal analysis brings to light a different behaviour for the seven elements tested.
Barium is a good paleonutritional indicator in this sample. The average values of the faunal groups increase from carnivores to herbivores as theoretically expected and, in general, the differences reach statistical significance. Although this element is generally considered subject to diagenetic alterations (Ahlgren et al. 1981, Badone & Farquhar 1982), some authors have proposed Barium as a better indicator than the more commonly used Strontium (Sillen & Kavangh 1982, Lambert et al. 1984) because of its larger atomic weight and radius.
Copper shows average contents in accordance with theoretical expectation in the faunal groups (decreasing from carnivores to herbivores), but statistical significance is reached in the Arene Candide - Neolithic subsample only.
Magnesium shows positive results for Arene Candide cave, while in the Uzzo sample a diagenetic change can be detected.
Previous work on Manganese has shown a high susceptibility to diagenesis for this element (Badone & Farquhar 1982, Lambert et al. 1982, 1983, 1984). For this reason, Mn was dosed here in order to control the behaviour of a certain diagenetic element. Mn confirms its extreme susceptibility to diagenesis: The differences among faunal groups were randomly distributed and the coefficients of variation (c.v.) were the highest observed here.
In spite of the fact that Strontium is generally considered a good paleonutritional indicator (Brown 1974, Schoeninger 1979, Sillen 1981, Lambert et al. 1984), it showed a poor value in this sample. While the average c.v. are close to the values found in living and past populations (ca 20%) (Schoeninger 1979, Sillen 1981, Sillen & Kavanagh 1982, Price 1985), a remarkable homogeneity can be noted in the faunal and human remains from all levels and all sites.
Vanadium shows very clear evidence of diagenetic change in the Uzzo sample, where a remarkable increase is observable in the Neolithic subsample.
Zinc behaves as a good dietary indicator in this sample and, together with Barium, discriminates clearly among faunal groups. The difference among average values (decreasing from carnivores to herbivores) are in accordance with theoretical previsions and, in general, statistically significant.

Figure 1 shows the results of the soil analysis.
With the exception of V, the elements analysed show a gradient decreasing with the distance from the bone for the first 10 cm. This finding indicates a similar behaviour (a loss from bone, according to Lambert et al. 1985) for the six elements, contrasting with the information obtained from faunal analyses (elements behave as differentially susceptible to diagenesis). A tentative explanation is that the loss could mask nutritional patterns for some elements only. In any case, this control, being

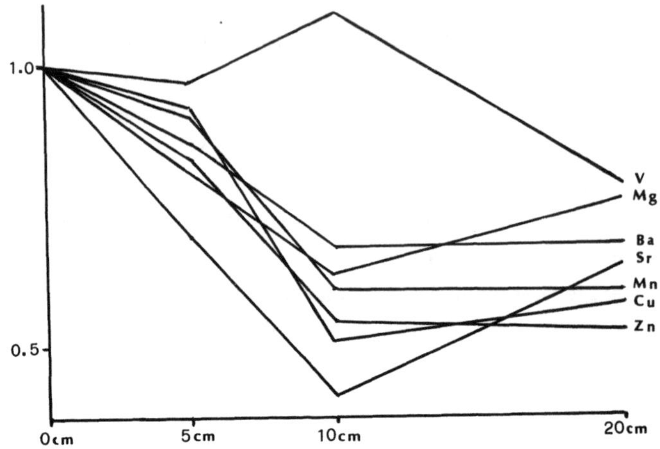


Figure 1. Results of the soil analysis. Values are expressed as a ratio to the amount detected at 0 cm.

site-specific and bone-specific, cannot provide results valid for the whole sample. The comparison between Arene Candide and Uzzo subsamples show that these two sites differ significantly for 3 elements (Mg: $t=4.38$ $P<.001$; V: $t=6.65$ $P<.001$; Zn: $t=2.85$ $P=.003$), pointing to a high geographic variability. It thus seems impossible to perform direct comparison among samples from different regions.

As reported in tables 3-5, the average Zinc content in the human remains is very high, even higher than in carnivores. Zinc is related to meat intake (Gilbert 1985), but it is very unlikely that the habits of human populations were more carnivorous than those of wild cats, foxes and some mustelids, represented in the "carnivores" subsamples. A more reasonable explanation is the possible consumption of marine molluscs, which contain very high amounts on Zn in the edible portion (Underwood 1977). The proximity to a rocky coast and the presence of shells in the deposit for both cave sites support this interpretation.

To test the above hypothesis directly, some samples were analysed for the stable isotopes ratio of Carbon and Nitrogen. The results are reported in Table 6.

The values found for Carbon isotopes lie in a very narrow range. No differences can be detected between humans and herbivores. On account of the environmental situation (proximity to the sea and presence of $C_3$ and $C_4$ plants), the $^{13}C/^{12}C$ ratio is not informative for the Mediterranean basin (Tauber 1981, 1986).

| | | $\delta^{13}C$ | | | $\delta^{15}N$ | |
|---|---|---|---|---|---|---|
| Subsample | n° | $\bar{x}$ | | n° | $\bar{x}$ | |
| Herbivores | 6 | -19.40 | 0.81 | 3 | + 4.18 | 0.71 |
| Humans | 7 | -19.74 | 0.73 | 4 | +12.89 | 6.73 |

Table 6. Results of the stable isotopes analysis. Ammonium sulphate ($\bar{x}$ = +1.71°/oo, $\sigma$ = 0.43, n=10) was used as standard for N dosage.

As regards the $\delta^{15}N$, the values measured in herbivores are in the range indicated by previous works (DeNiro & Epstein 1981, Schoeninger & DeNiro 1984, DeNiro 1985). The $\delta^{15}N$ in human bones is higher than that in the herbivores. The values are indicative of both a carnivorous diet and an appreciable intake of marine foodstuff. The subject AC XI has a $\delta^{15}N$=+22.0°/oo, similar to that found in marine mammals from other sites (Schoeninger & DeNiro 1984, Walter & DeNiro 1984). AC XI also shows the highest Zn content (413.2 ppm). Spearman's correlation coefficient between $\delta^{15}N$ values and Zn content in this subsample is significant ($r_s$=0.852; .05>P>.01). This seems to confirm that a portion of the Zn detected in human bones is of marine origin.
A detailed multivariate statistical analysis of the paleonutritional reconstruction for Arene Candide and Uzzo caves, which is beyond the aim of this paper, can be found in Francalacci (1987).

## Conclusions

The high intra-individual variability of trace element contents seems to be a major problem. It could be caused by several factors, such as metabolic differences in various skeletal districts, dishomogeneity of the bone tissue and, after death, local differential diagenetic changes (due to the action of microorganisms and/or to the geochemical environment). In order to avoid misinterpretation of the results, this variability should be limited by choosing the same skeletal element and standardizing the drawing procedure, or, even better, quantified by analysing many different bones for each individual.
Where it is impossible to perform repeated tests, as in the case of incomplete skeletons or very precious remains, individual dietary reconstruction should not be attempted by trace element analysis.
Another important problem is the recognition and evaluation of diagenetic changes. The behaviour of an element in respect to diagenesis proved unpredictable *a priori*.

In this sample, for instance, Ba behaves as a good paleonutritional indicator, while Sr does not distinguish between herbivores and carnivores.

In general, a multielementary approach is preferable, at least in a selected subsample. It should be accompanied by appropriate controls testing the presence and effects of diagenetic alterations, so as to choose the more reliable elements to be dosed in the whole sample.

The results of this research indicate that faunal analysis seems to be the most effective control for paleonutritional studies. If an element allows the correct discrimination among faunal subsamples of known diet, it should also be suitable for nutritional inference in human remains from the same environmental and depositional context. Faunal analysis cannot exclude the presence of diagenesis, as in cases of leaching or enrichment proportional to the ante mortem levels. This can explain the contrasting results obtained from the two controls: A general leaching of all elements is shown by soil analysis, while a differential diagenetic pattern appears in faunal analysis. In any case, this is not a real problem when the dietary reconstruction is not based on absolute amounts of trace elements, but is performed using faunal reference samples of known diet. Moreover, on account of the high geological variability in the distribution of trace elements, only faunal analysis allows comparison among samples from different regions.

Preliminary results on parallel dosage of trace elements and stable isotopes are very promising. These are two independent approaches from both theoretical and technical standpoints (different substrata and methods). They are partly complementary and can provide a reciprocal control of results (e.g. $\delta^{15}N$ and Zn as indicators of the dietary incidence of marine foodstuff).

In conclusion, in order to obtain reliable results from trace element analysis, the following suggestions are appropriate:

- a preliminary set of tests aimed at evaluating intra-individual variability and an appropriate choice of bones, tissues and elements;
- instrumental controls with appropriate standards and the use of the same instrument for each sample;
- parallel dosages on soil from the same layer;
- parallel dosages on faunal remains of known diet;
- parallel stable isotopes dosages.

If most of the above mentioned suggestions are followed, trace element analysis is a reliable dietary indicator and, especially when applied together with other investigations (e.g. dental microwear and paleopathology), a valuable tool for the reconstruction of subsistence patterns of ancient human populations.

Acknowledgements

We thank the staff of ENEL-DCO Laboratory, Piacenza (Dr. Bettinelli, Mr. Baroni and Mrs. Pastorelli) for helping in ICP analysis, the staff of the Isotopen Fysica Laboratorium (University of Groningen), directed by Prof. Mook, for helping in stable isotope analysis and Dr. Van Klinken (I.F. Lab., Groningen) for advice on collagen extraction and combustion. We are also indebted to Prof. Zamboni (Dept. of Chemistry, University of Pisa) for advice on the preparation of the sample. Thanks are due to the Soprintendenza Archeologica della Liguria, The Museo di Pegli (Dr. Isetti), the Istituto Italiano di Paleontologia Umana (Dr. Segre) and the Museo Pigorini (Dr. Piperno) for permission to sample the Arene Candide and Uzzo materials.

References

Ahlgren L, Christoffersson JO, Mattsson S (1981) Lead and barium in archaeological Roman skeletons measured by nondestructive X-ray fluorescence analysis. In: Smith DJ, Barrett CS, Leyden DE, Predecki PK (eds) Advances in X-ray analysis. Plenum, New York, 24: 377. Quoted in Lambert et al. (1984).

Badone E, Farquhar RM (1982) Application of neutron activation analysis to the study of element concentration and exchange in fossil bone. J Radioanal Chem 69: 291-311.

Bietti A (1987) Some remarks on the new radiocarbon dates from the Arene Candide cave (Savona-Italy). Hum Evol 2: (in press).

Brown AB (1974) Bone strontium as a dietary indicator in human skeletal populations. Contr Geol 13: 47-48.

Brown AB, Blakely RL (1985) Biocultural adaptation as reflected in trace elements distribution. J Hum Evol 14: 461-468.

DeNiro MJ (1985) Postmortem preservation and alteration of in vivo bone collagen isotope ratios in relation to paleodietary reconstruction. Nature 317: 806-809.

DeNiro MJ, Epstein S (1981) Influence of diet on the distribution of nitrogen isotopes in animals. Geochim Cosmochim Acta 45: 341-351.

Francalacci P (1987) Oligoelementi e paleonutrizione: Aspetti metodologici e applicativi in due giacimenti italiani (Grotta delle Arene Candide e Grotta dell'Uzzo). Ph.D. thesis, University of Pisa.

Gilbert RJ Jr (1985) Stress, paleonutrition and trace elements. In: Gilbert RJ Jr, Mielke JH (eds) The analysis of prehistoric diets. Academic Press, Orlando, pp 339-357.

Grupe G, Piepenbrink H (1987) Trace element contaminations in excavated human bones by microorganisms. See this book, pp 103-112

Herrmann B, Grupe G (1987) Trace element content in cremated human remains. See this book, pp 91-101

Klepinger LL, Kuhn JK, Williams WS (1986) An elemental analysis of archaeological bone from Sicily as a test of predictability of diagenetic change. Am J Phys Anthrop 70: 325-331.

Lambert JB, Simpson SV, Buikstra JE, Hanson D (1983) Electron microprobe analysis of elemental distribution in excavated human femurs. Am J Phys Anthrop 62: 409-423.

Lambert JB, Simpson SV, Szpunar CB, Buikstra JE (1984) Ancient human diet from inorganic analysis of bone. A Chem Res 17: 298-305.

Lambert JB, Simpson SV, Szpunar CB, Buikstra JE (1985) Bone diagenesis and dietary analysis. J Hum Evol 14: 477-482.

Lambert JB, Vlasak SM, Thometz AC, Buikstra JE (1982) A comparative analysis of ribs and femurs in Woodland populations. Am J Phys Anthrop 59: 289-294.

Maggi R (1977) Lo strato a ceramiche graffite delle Arene Candide. Preist Alp 13: 205-211.

Meulengracht A, McGovern P, Lawn B (1981) University of Pennsylvania Radiocarbon dates. Radiocarbon 23: 227-240.

Piperno M (1985) Some C dates for the paleoeconomical evidence from the Holocene levels of the Uzzo cave, Sicily. Proc 3rd Conf Italian Archaeol, Cambridge, 6/8-1 1984. B.A.R. Int Ser 244: 83-86.

Price TD (1985) Late archaic subsistence in the Midwestern United States. J Hum Evol 14: 499-459.

Price TD, Connor M, Chase EP (1985) Bone chemistry and the reconstruction of diet: Strontium discrimination in white-tailed deer. J Archaeol Sci 12: 419-442.

Schoeninger MJ (1979) Diet and status at Chalcatzingo: Some empirical and technical aspects of strontium analysis. Am J Phys Anthrop 51: 295-310.

Schoeninger MJ, DeNiro MJ (1984) Nitrogen and carbon isotopic composition of bone collagen from marine and terrestrial animals. Geochim Cosmochim Acta 48: 625-639.

Sillen A (1981) Strontium and diet at Hayonim cave. Am J Phys Anthrop 56: 131-137.

Sillen A, Kavanagh M (1982) Strontium and paleodietary research: A review. Yearb Phys Anthrop 25: 67-90.

Szpunar CB, Lambert JB, Buikstra JE (1978) Analysis of excavated bones by atomic absorption. Am J Phys Anthrop 48: 199-202.

Tauber H (1983) C dating of human beings in relation to dietary habits. Pact J 8: 365-375.

Tauber H (1986) Analysis of stable isotopes in prehistoric populations. Mitt Berl Ges Anthrop Ethn Urgesch7: 31-38.

Tine S (1974) Il Neolitico e l'età del Bronzo della Liguria alla luce delle recenti scoperte. Atti XVI Riun Sc Ist It Preist Protost 16: 37-54.

Underwood EJ (1977) Trace elements in human and animal nutrition, 4th edn. Academic Press, New York.

Walker PL, DeNiro MJ (1986) Stable Nitrogen and Carbon isotope ratios in bone collagen as indices of prehistoric dietary dependence on marine and terrestrial resources in Southern California. Am J Phys Anthrop 71: 51-61.

# DISCRIMINATION FACTORS ON DIFFERENT TROPHIC LEVELS IN RELATION TO THE TRACE ELEMENT CONTENT IN HUMAN BONES

Lex Runia
Albert Egges van Giffen Instituut voor Prae- en Protohistorie
University of Amsterdam
Singel 453
1012 WP Amsterdam, Netherlands

## Introduction

The concentration of strontium and other trace elements in bones of living humans and animals reflect the levels in the environment (Underwood 1977, Rheingold et al. 1983, Price et al. 1985b). Historic and prehistoric bones may thus contain information on past environments. The chemical composition of bones that have been buried for a long period may, however, have changed by diagenetic processes, which result in the loss of information. The concentrations of some of the trace elements in bones appear to be fairly stable and may remain almost unchanged for very long periods of contact between bone and soil.

The study of diagenetic changes of trace element concentrations in bones is essential for the study of prehistoric bones and the interpretation of their trace element levels, but this aspect will not be discussed in detail here. Because every soil condition and diagenetic environment is unique, it is hazardous to generalize on the behaviour of trace elements in buried bones. Bone chemistry studies have, however, indicated that the concentrations of a few trace elements, e.g. strontium and zinc, are comparatively unsusceptible to diagenetic influences (Lambert et al. 1979, 1982, Sillen & Kavanagh 1982, Gilbert 1985). It appears therefore justified to restrict a discussion about the relation between trace element concentrations in the environment and in human bones to these two elements. As will be shown in the following sections, even for these comparatively well-known elements many problems remain present.

Trace elements can enter the body either through the skin, the lungs or the gastro-intestinal tract. The uptake of elements from the skin and the lungs can be of importance in certain cases: Uptake of elements from inhaled dust or fumes may be significant in mining and industrial environments (Holdgate 1979). Uptake via the skin can also occur in such environments, but also when cosmetics or paints are applied to the skin (Nriagu 1983). In most (pre)historic contexts, these routes of entry are of minor importance in comparison to the uptake of elements from the intestines. Foods and, to a lesser extent, water, can therefore be considered the main sources of strontium and zinc.

The discrimination processes which determine the trace element content in (living) human bones can be devided in (1) processes operating within the body, i.e. after the food has entered the body through the mouth, and (2) the processes which determine the concentrations of the elements in the food. In the following sections, these processes will be discussed. As stated above, processes (3) which influence bone trace element levels ante mortem, will not be discussed. The focus of the discussion will be on Sr, because the relationship between environmental and bone concentrations for this non-essential element are more straightforward than for the essential element Zn.

## Discrimination factors within the human body

Once foods have been ingested, part of the elements will pass the wall of the intestines. The elements are transported through the body in the blood stream, and may be (temporarily) deposited in body tissues, such as bone. Elements can leave the body either after filtering the blood in the kidneys or in another form, such as hair and nails. Women may loose elements in the foetus and in mother's milk.

The turn-over of Sr in the body has been extensively studied and the relationship between dietary levels and concentrations in bones is therefore comparatively well documented (Comar et al. 1957, Comar 1967). The relationship between Zn levels in food and bone are much more complicated, because Zn plays a role in a number of enzymes (Prasad 1976, Underwood 1977).

Because of its chemical similarity to calcium, the behaviour of Sr in the body resembles Ca. In consequence, the relationship between Sr in food and bones is usually expressed as the

Observed Ratio = (Sr/Ca ratio in bone)/(Sr/Ca ratio in food)

as defined by Comar et al. (1957). The OR is the product of all discrimation processes at various sites, such as kidneys and mammary glands in the body (Discrimination factors, DF):

$$OR = DF_1 \times DF_2 \times .... \times DF_n$$

The over-all OR bone-diet for the Sr/Ca system is qualitatively and quantitavely reasonably well known, most mammals on non-milk diets showing values close to 0.25 (Comar et al. 1957, Comar 1967, Sillen & Kavanagh 1982). Neonatal children do not discriminate against Sr, but within a few months after birth adult OR values appear to be present (Lengemann 1963). The effects of certain dietary components on OR values are, however, not well known in a quantitative sense. Lactose, for instance, is known to stimulate the uptake of Ca from the intestinal tract and to decrease the discrimination against Sr (Wasserman et al. 1957, Dupuis & Fournier 1963), whereas phytate may inhibit the uptake of Zn and probably other elements (van Barneveld 1983). The effects of these components in low concentrations and in combination with other foods are, however, not known in detail. Still another problem is formed by the effects of pregnancy and lactation (Sillen & Kavanagh 1982, Price et al. 1985a).

No OR values for the Zn/Ca system can be established. Zinc can be stored temporarily in the bone, but may be mobilized again under severe Zn stress (Underwood 1977). The active treatment of Zn thus masks a clear relationship between dietary and bone Zn concentrations. However, bone Zn levels roughly correspond to dietary levels as may be inferred from different bone Zn in herbivores, omnivores and carnivores (Rheingold et al. 1983).

## Strontium and zinc in foods: general aspects

Because the relation between Sr/Ca ratios in diet and bone are comparatively well known, the dietary Sr/Ca ratios and the discrimination processes which determine them are of primary interest. A generalized picture of the Sr/Ca ratios in dietary components from different trophic levels is given in fig. 1. Zinc levels (Zn/Ca ratios) in the food chain show an opposite trend (tab. 1). Obviously, the actual situation is much more complex.

The Sr/Ca ratio of a diet depends on the Sr/Ca ratios of its components. The effect of a food item on the overall Sr/Ca ratio of a diet can be seen as the result of three factors: (1) its Ca content, (2) the Sr/Ca ratio in relation to those of the other components of the diet, and (3) the amount that is consumed or the relative portion of food item in the diet (Runia 1987c). It is obvious that the Sr/Ca ratios in food items with high Ca contents, such as milk and leafy vegetables (tab. 1) will to a major extent govern the Sr/Ca ratio of a diet. This has been illustrated graphically for milk (Runia 1987a).

Table 1. Mean energy content, and mean concentrations of Ca and Zn per unit of energy in groups of modern food items. Sr/Ca ratios are approximations of values relative to Sr/Ca ratios in green, young plant shoots. Calculated from data in Paul & Southgate (1978) and various other sources (collected in Runia, 1987b,c).

| food | MJ/kg | --mg per MJ-- Ca | Zn$ | mg Zn/g Ca$ | relative Sr/Ca ratio |
|---|---|---|---|---|---|
| leafy vegetables | 0.8 | 1420 | 5.1 | 6 | 1 |
| roots | 1.3 | 450 | 2.6 | 8 | 1.6 |
| cereals | 13.5 | 30 | 2.2 | 86 | 1.5-2 |
| nuts | 18.9 | 60 | 1.4 | 39 | ? |
| fruits | 14.4 | 200 | 0.7 | 10 | ? |
| | | | | | |
| milk | 2.7 | 440 | 1.3 | 3 | 0.1 |
| meat# | 13.5 | 5 | 1.9§ | 342§ | 0.25 |
| sea fish | 5.7 | 490 | 0.9§ | 21§ | - |
| marine shell fish | 3.9 | 410 | 35 | 73 | - |

$calculated from primary data; no. of items with known Zn content smaller than for Ca contents

#dressed carcasses

§value for sea and fresh water fish

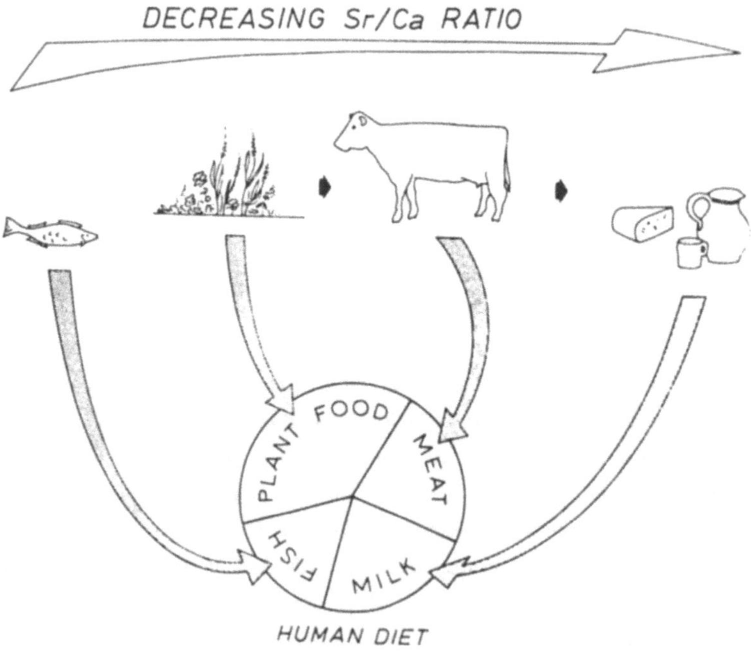


Figure 1. Components of human diet can be taken from different trophic levels. Discrimination processes cause a decrease of Sr/Ca ratios in the higher trophic levels.

The uptake of Zn from the intestines is fairly complicated and depends on a number of factors, such as Zn status of the individual, the presence of uptake of enhancing of inhibiting compounds, and the presence of other cations such as Fe, Cu and Ca (Underwood 1977, Jonxis 1983). In general, Zn is better taken up from animal foods than from vegetable foods (Underwood 1977). This may partly be the result of the high Zn/Ca ratios in the meat (tab. 1).

Descriptions of dietary compositions of diets can be performed by expressing the amounts of consumed food items in kilogram wet weight. It may, however, be preferred to express diets in terms of amounts of energy. In my view, this is justified because the human energy demands largely determine the amounts of food items consumed. The use of energy units may, moreover, facilitate a quantitative approach of diet and the comparison with energy costs of purchasing food. When diets are being discussed, it should be clearly indicated which units are used.

## Components of human diet

On the bill of fare and foods consumed by human beings a very wide variety of items is listed. In anthropological and historical literature consumption of almost anything, ranging from clay to toxic plants and from insects to human brains, is reported (Brothwell & Brothwell 1969, Hayden 1981).

The actual composition of a diet depends on a number of factors, such as environmental conditions, including climate and season, cultural and agricultural factors, and individual factors, which include socio-economic aspects such as sex, age, wealth and social ranking. These aspects are not of interest here, but we have to be aware of the wide range of potential food items, which may be taken from any trophic level in the food chain.

A distinction between a terrestrial and a marine food chain can be made . Coast dwelling cultures may for a large part depend on the marine foods fish, shell fish and seaweeds, and on meat and eggs produced by animals feeding on marine food sources. Terrestrial foods comprise well-known products from agriculture or gathered from the wild (cereals, root crops, nuts, meat, milk) but also other items such as mushrooms, bird and reptile eggs, insects and snails. As stated above, each component exerts a certain influence on the diets' over-all trace element composition. A discussion on trace element concentrations in (prehistoric) human bones can, therefore, only be attempted when the range of potential food items, and their respective trace element levels, of the (archaeological) context involved is known.

## The marine food chain: discrimination factors

Sea water is characterized by comparatively high concentrations of many elements (Weast 1973), and it is thus to be expected that marine foods have high trace element levels. The discrimination factors determining the Sr/Ca ratios and Zn contents of marine foods are not very well knwon. Aquatic plants may discriminate Sr or against Ca (Isermann 1981). Seaweeds may be part of human and animal diets, and can also indirectly, when used as fertilizer, supply elements to the terrestrial food chain. Seaweeds probably have high trace element contents (Luten 1983), but no data on Sr/Ca ratios are available.

Fish can take up elements from the water via gills and skin, and from food in the intestines, the first route being the most important (Chipman 1966). Values for the Sr/Ca OR tissue-water range from 0.2 to 0.8, and marine shellfish appear to have OR shell-water values in the same range (Rosenthal et al. 1970). There are no data on Sr/Ca ratios in the soft, edible tissues of shell fish, but it is possible that differences between edible parts and shells exist.

In sum, the discrimination processes in the sea are quantitatively not well documented. It is, however, likely that all marine foods have Sr/Ca ratios that are higher than those in terrestrial foods, and that Zn contents may also be comparatively high.

## The terrestrial food chain: discrimination factors

Much of the research that has been carried out because of the dangers of radiostrontium was focussed on the terrestrial food chain. The discrimination against Sr on certain trophic levels is, therefore, comparatively well knwon. Values for the OR bone-diet and OR milk-diet have been established for a number of mammalian species, and average 0.2-0.25 and 0.1, respectively (Comar et al. 1957, Comar 1967, Sillen & Kavanagh 1982).

Data on other trophic levels and on Sr/Ca ratios in large groups of terrestrial foods are scarce. No information on Sr/Ca ratios in bird and reptile meat and eggs, invertebrates (insects, snails) and lower plants (mushrooms) are available though these may be important foods.

At the basis of terrestrial food chains are plants. Plants and plant parts appear to show a wide range of Sr/Ca ratios (Isermann 1981, Runia 1987b) and these differences will be reflected in higher trophic levels. Knowledge of Sr/Ca ratios in plants and plant parts is, therefore, essential for the interpretation of Sr levels in human bones.

Soil particles ingested by man and animals may be important sources of macro and trace elements (Thornton, this volume). All discriminating factors in the food chain are obviously evaded by this route of entry. Some qualitative data on the intake of soil are available (Hayden 1981, Thornton, this volume), but at this moment it is not possible to assess the importance of this process in the supply of Ca and trace elements. Coprolites of humans and various animals may provide some information regarding the ingestion of soil in (pre)historic times. It is, however, not customary to determine the soil or ash content of coprolites (Fry 1985, Paap & Troostheide, pers. comm.).

## Vegetable material

In paleodietary studies it is generally assumed that within vegetable tissues no differences in Sr/Ca ratios exist (Sillen & Kavanagh 1982, Price et al. 1985b). This is, however, a simplification of the actual complex situation (Runia 1987b,c). As has been pointed out by Comar et al (1957), there was (and still is) insufficient knowledge on the Sr/Ca ratios in the storage organs of plants. Because storage organs are main sources of vegetable food for humans and certain animals, this may cause problems in the interpretation of Sr concentrations in human bones and when these are compared with Sr levels in bones of herbivorous baseline species (Runia 1987a).

Because of the different nature of discrimination processes in plants, the term Distribution Ratio has been proposed: DR = (Sr/Ca ratio in x)/(Sr/Ca ratio in y), x and y being plant parts or growth media (Runia 1987b).

Cereals are staple foods in the Old World. Although they have low Ca contents (tab. 1), the large proportion of cereals in many diets makes the discrimination against Sr in cereals of interest for paleodietary studies. Published data show that Sr/Ca ratios in grains differ from those in (young) shoots and straw (Smith 1971b) while considerable differences exist between various cereal species and varieties (Smith et al. 1963).

Samples of various cereal species and other vegetable matter have been analysed in order to get some quantitative knowledge on the discrimination processes in cereals (Runia 1987a,c). Samples were taken from three small trial fields, one of which was on unfertilized sandy soil (plot a), one on fertilized sandy soil (plot b) and the third on fertilized river clay (plot c). Three primitive grain species were sampled: emmer

(Triticum dicoccum), einkorn (Triticum monococcum) and hulled barley (Hordeum distichum). The emmer and einkorn were grown as a mixed crop. Concentrations of Sr, Ca and Zn were determined after wet digestion using ICP-AES and AAS. For a detailed description of the experiment reference is made to Runia (1987c).

The results of the experiment show that large differences exist between the Sr/Ca ratios of the cereals. Sr/Ca ratios of Triticum dicoccum and T. monococcum grown as a mixed crop (three cases) differed by factors of 1.3, 3.5 and 2.9 for the plots a, b and c, respectively. These data suggest that cereal species react differently on variations in growing conditions.

Table 2. Sr/Ca Distribution Ratios for grains of some cereal species. The mean Sr/Ca ratios in young, green plants from the same soils are used as reference: DR = (Sr/Ca ratio in grain)/(reference Sr/Ca ratio). For details see Runia (1987c).

| species | no. of cases | Distribution Ratio | |
|---|---|---|---|
| | | mean | SD |
| Triticum dicoccum | 4 | 1.5 | 0.3 |
| Triticum monococcum | 4 | 3.6 | 1.7 |
| Hordeum distichum | 2 | 1.3 | |
| Triticum aestivum 'Gelderse Ris' | 2 | 2.0 | |

Shoots of cereals and pooled samples of green weds from the plots were analysed as reference material. Relatively young, green plant parts and shoots appear to have comparable Sr/Ca ratios, which probably reflect the ratio in which Sr and Ca is available in the soil. This finding agrees with the results of numerous greenhouse experiments (for instance, Collander 1941, values compiled in Runia 1987b). It can be expected that herbivores of the grazer type, such as cattle, will ingest plant material with Sr/Ca ratios close to those in the reference material.

The relations between Sr/Ca ratios in the reference material and in the cereals are presented in tab. 2. The results suggest that in comparison to Triticum dicoccum, the Sr/Ca ratios in T. monococcum grains depend more on the growing conditions. On average, the grains have Sr/Ca ratios that are $2.2 \pm 1.4$ (n=12) times higher than in the reference material. The vegetable part of many human diets may, therefore, have Sr/Ca ratios which are distinctly higher than those in the animal fodder in the same region.

Zinc concentrations and Zn/Ca ratios in grains of various cereals show differences, but are all much higher than in the young, green plant parts. A stock of the essential element Zn is apparently incorporated in the grains. Within the grains, Zn is predominantly present in the bran as is phytate (van Barneveld 1983), which may reduce the amount of Zn that can be taken up in the intestines.

An interesting aspect of cereals is the difference between bran and endosperm. Bartlett & Gunn (1962) found their Sr/Ca ratios to differ by a factor of 0.7. Wholemeal and white breads made from the same stock of grain will thus probably show different Sr/Ca ratios.

Roots of various plants are edible and may be important as foods for man and animals. Published shoot/root distribution ratios suggest that roots have relatively high Sr/Ca ratios (for instance, Menzel & Heald 1955, values compiled in Runia 1987c). Levels of other elements, such as Zn, may be comparatively high in roots.

Roots from several plant species and weeds cultivated on plot a and b were sampled and analysed (tab. 3).

Table 3. Sr/Ca Distribution Ratio for primary roots of some plant species: DR = (Sr/Ca ratio in roots)/(reference Sr/Ca ratio). The Sr/Ca ratios in young, green plant material from the same soils are used as reference. For details see Runia (1987c).

| species | Distribution Ratio |
|---|---|
| *Cannabis sativa* | 1.5 |
| *C. sativa* | 1.7 |
| *Fagopyrum esculentum* | 1.3 |
| *Taraxacum* sp. | 2.0 |
| various weeds | 1.6 |
| mean (n=5) | 1.6 $\pm$0.2 |

The results of our experiment agree with published data in that the Sr/Ca ratios in roots were distinctly higher than in young, green above-ground plant parts. Roots have high Ca concentrations in comparison to cereals, and roots in diets will, therefore, tend to increase dietary Sr/Ca ratios even if they supply only a minor part of the energy.

Data on Sr/Ca ratios in other important vegetable foods are scarce. Smith (1971a) analysed roots, leaves and nuts of the Brazil Nut; this species may, however, possess a deviant physiology (Runia 1987b). Useful data on the relationship between Sr/Ca ratios in nuts and green, young plant material from the same soil are not available as is the case for fruits (berries, apples and others). Leafy vegetables, which usually are young, green plant parts, will probably show Sr/Ca ratios close to the value for available Sr and Ca in the soil. Leafy vegetables in human diets will tend to bring its Sr/Ca ratio closer to the Sr/Ca ratio of the material consumed by herbivores.

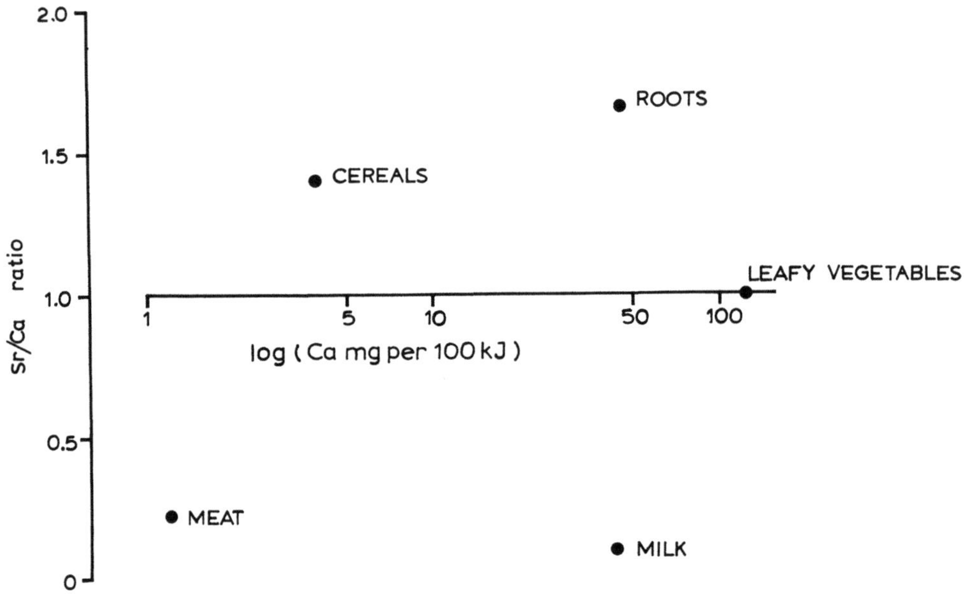


Figure 2. Mean Ca concentrations (log scale) and relative Sr/Ca ratios (=(Sr/Ca ratio of food item)/(reference Sr/Ca ratio); vertical axis) in some important groups of food items. The Sr/Ca ratios of young, green plant material (e.g. leafy vegetables) is used as a reference value for the relative Sr/Ca ratios.

## Food selection

The principle of food selection by humans stands implicitely at the basis of paleodietary studies. The food selection by animals receives, however, little attention. Toots & Voorhies (1965) studied bones of, potentially, grazing or browsing herbivores, but since then animals are usually roughly devided in three groups: herbivores, carnivores and omnivores. In reality, every animal species and probably every individual animal has its own preferred food, whether vegetable or animal. Because of the discrimination processes in plants, the selection of specific plants and plant parts by herbivores and omnivores will result in different levels of trace elements in animals of various species and even of individuals of the same species.These differences will be reflected on higher levels in the food chain, i.e. in carnivores and omnivores, including humans.

An example of the effect of food selection on bone trace element levels is the mean level of Sr in pig bones compared with cattle bone Sr levels from Bronze Age sites in West-Friesland (Runia 1987c). One should expect the strict herbivorous cattle to have higher Sr levels than the omnivorous pigs. Pigs have, however, the higher Sr concentrations of the two, which might be explained by them consuming plant roots, which have comparatively high Sr/Ca ratios (tab. 1,3). Pork and beef from animals raised at the same site in such cases would differ Sr/Ca ratios, which might affect the over-all Sr/Ca ratios in 'high' and 'low' status diets.

The effects of food selection by animals are insufficiently known. In order to be able to interpret Sr and Zn levels in prehistoric and animal bones, reference data from present-day ecosystems should be present. Though some studies on this aspect have been carried out (Price et al. 1985a, Schoeninger 1985), detailed research on Sr and Zn levels in vegetable material and in various animal species from a restricted area remains to be done.

## Conclusions

Discrimination factors causing different Sr/Ca ratios and Zn levels in components of diets are of prime interest for the interpretation of Sr and Zn levels in human bones. Dividing vegetable foods in various groups and determining the mean Ca contents and relative Sr/Ca ratios in these groups (fig. 2) gives a better background to the interpretation of human bone Sr levels than the simple scheme of fig. 1. The composition of real prehistoric diets is, obviously, still much more complex, and the interpretation of trace element levels in bones can only be attempted when the range of potential food items and their respective trace element levels are known. The comparison with herbivorous baseline species also depends on such information: Grazers and browsers may consume food with different trace element levels, while the vegetable component of human diet probably has still another chemical composition.

Though discrimination factors on Sr and Ca in some sections of terrestrial food chains are fairly well understood, the discrimination in other sections is quantitatively and qualitatively still poorly documented. Much of the information present is derived from laboratory experiments which fail to take the complex interactions in actual field situations into account. For the essential element Zn, the relationships between concentrations on various trophic levels may be even more complex. It is, therefore, essential for the interpretation of trace element levels in human and animal bones from archaeological contexts that reference material from comparable, present-day ecosystems becomes available. This reference material should comprise all sorts of vegetable and animal foods, and animal bones from a restricted area. In many instances, this material will have to be collected and analysed by chemo-archaeologists themselves because their questions will often be uninteresting for other fields of research.

## References

Barneveld AA van (1983) De invloed van fytaatin de voeding: een overzicht. Voeding 44: 51-58.

Bartlett BO, Gunn KB (1962) Discrimination between stable strontium and calcium in cereal grain. Agr Res Counc Radiob Lab Ann Rep 1961-62: 77-78.

Brothwell DR, Brothwell P (1969) Food in antiquity. Thames and Hudson, London.

Chipman WA (1966) Food chains in the sea. In: Russel RS (ed) Radioactivity and human diet. Pergamon Press, Oxford. p421.

Collander R (1941) Selective absorption of cations by higher plants. Plant Phys 16: 691-720.

Comar CL (1967) Some principles of strontium metabolism: Implications, applications, limitations. In: Lenihan JMA, Loutit JF, Martin JH (eds) Strontium metabolism. Academic Press, London. p 17.

Comar CL, Russell RS, Wasserman RH (1957) Strontium-calcium movement from soil to man. Science 126: 485-492.

Dupuis Y, Fournier P (1963) Lactose and the absorption of calcium and strontium. In: Wasserman RH (ed) The transfer of calcium and strontium across biological membranes. Academic Press, New York. p 277.

Fry GF (1985) Analysis of fecal material. In: Gilbert RI, Mielke JH (eds) The analysis of prehistoric diets. Academic Press, Orlando. p127.

Gilbert RI (1985) Stress, paleonutrition and trace elements. In: Gilbert RI, Mielke JH (eds) The analysis of prehistoric diets. Academic Press, Orlando. p 339.

Hayden B (1981) Subsistence and ecological adaptations of modern hunter/gatherers. In: Harding RSO, Teleki G (eds) Omnivorous primates. Columbia Univ Press, New York. p 344.

Holdgate MW (1979) A perspective of environmental pollution. Cambridge Univ Press, Cambridge.

Iserman K (1981) Uptake of stable strontium by plants and effects on plant growth. In: Skoryna SC (ed) Handbook of stable strontium. Plenum Press, New York. p 65.

Jonxis JHP (1983) Sporenelementen. Voeding 44: 266-273.

Lambert JB, Szpunar CB, Buikstra JE (1979) Chemical analysis of excavated human bones from Middle and Late Woodland sites. Archaeometry 21: 115-129.

Lambert JB, Vlasak SM, Thometz AC, Buikstra JE (1982) A comparative study of the chemical analysis of ribs and femurs in Woodland populations. Amer J Phys Anthrop 59: 289-294.

Lengemann FW (1963) Over-all aspects of calcium and strontium absorption. In: Wasserman RH (ed) The transfer of calcium and strontium across biological membranes. Academic Press, New York. p 85.

Luten JB (1983) Spoorelementen in voor de consumptie bestemde zeewieren. Voeding 44: 232-236.

Menzel RG, Heald WR (1955) Distribution of potassium, rubidium, cesium, calcium and strontium within plants grown in nutrient solutions. Soil Sci 80: 287-293.

Nriagu JO (1983) Lead and lead poisoning in classical antiquity. John Wiley, New York.

Paul AA, Southgate DAT (1978) McCance & Widdowson's The composition of foods, 4th ed. HMSO, London; Elsevier/Norht-Holland, Amsterdam.

Prasad AS (ed) (1976) Trace elements in human health and disease, vol. 1: Zinc and copper. Academic Press, New York.

Price TD, Connor M, Parsen JD (1985a) Bone chemistry and the reconstruction of diet: strontium discrimination in white-tailed deer. J Archaeol Sci 12: 419-442.

Price TD, Schoeninger MJ, Armelagos GJ (1985b) Bone chemistry and past behaviour: an overview. J Hum Evol 14: 419-447.

Rheingold AL, Hues S, Cohen MN (1983) Strontium and zinc content in bones as an indication of diet. J Chem Educ 60: 233-234.

Rosenthal HL, Cochran OA, Eves MM (1970) Common strontium content concentration of mineralized tissues from marine and sweet water animals. Comp Biochem Phys 32: 445-450.

Runia LT (1987a) Analysis of bone from the Bronze Age site Bovenkarspel-Het Valkje, the Netherlands: a preliminary report. Archaeometry 29. (in press)

Runia LT (1987b) Strontium and calcium distribution in plants: effect on paleodietary studies. J Archaeol Sci 14. (in press)

Runia LT (1987c) The chemical analysis of prehistoric bones. A paleodietary and eco-archaeological study of Bronze Age West-Friesland. BAR, Oxford (BAR Int. Series). (in press).

Schoeninger MJ (1982) Diet and the evolution of modern human form in the Middle East. Amer J Phys Anthrop 58: 37-52.

Schoeninger MJ (1985) Trophic level effects on $^{15}N/^{14}N$ and $^{13}C/^{12}C$ ratios in bone collagen and strontium levels in bone mineral. J Hum Evol 14: 515-525.

Sillen A (1981) Strontium and diet at Hayonim cave, Israel: an evaluation of the strontium/calcium technique for investigating prehistoric diets. PhD diss., Univ. of Pennsylvania.

Sillen A, Kavanagh M (1982) Strontium and paleodietary research: a review. Yearb Phys Anthrop 25: 67-90.

Smith KA (1971a) The comparative uptake and translocation by plants of calcium, strontium, barium and radium. I. Bertholetta excelsa (Brazil Nut Tree). Plant Soil 34: 369-379.

Smith KA (1971b) The comparative uptake and translocation by plants of calcium, strontium, barium and radium. II. Triticum vulgare (wheat). Plant Soil 34: 643-651.

Smith LH, Rasmussen DC, Myers WM (1963) Influence of genotype upon relationship of strontium-89 to calcium in grain of barley and wheat. Crop Sci 3: 386-389.

Thornton I (this volume) Soil features and human health.

Toots H, Voorhies MR (1965) Strontium in fossil bones and the reconstruction of food chains. Science 149: 854-855.

Underwood EJ (1977) Trace elements in human and animal nutrition. Academic Press, New York.

Wasserman RH, Comar CL, Papadopoulou D (1957) Dietary calcium levels and retention of radiostrontium in the growing rat. Science 126: 1180-1182.

Weast RC (ed) (1973) Handbook of chemistry and physics, 54th ed. CRC Press, Cleveland.

# TRACE ELEMENTS IN HUMAN TEETH

Theya Molleson
Department of Paleontology
British Museum (Natural History)
London SW7 5BD, UK

## Introduction

Trace elements may be incorporated into the mineral structure of teeth, during the formation of a tooth, during the life-time of the individual and, in the case of material buried in the ground, after death. Much the same range of trace elements may be acquired by the tooth at any of these times and the distribution in the tooth of the elements will not necessarily differ. Yet the implications of trace element acquisition at different times in the life history of a tooth can differ enormously as regards the diet, health or burial conditions of the individual. It is only by a careful and detailed analysis of the distribution patterns of a suite of trace elements within the tooth, and between different teeth of the one individual that there is any possibility of extricating, one from another, the elemental uptake of different periods in the life of the individual.

## Trace element sources

The trace elements that are most commonly identified when teeth are analysed are shown in Table 1. Fluorine may be included here in that the uptake and incorporation of fluorine in the tooth structure follows the same pattern as do the trace elements. Uranium is frequently found in teeth that have been buried. Again the incorporation is similar to that of trace elements.

Most trace elements find their way into the human system via drinking water; and considerable variation in concentration of different elements in teeth of individuals from populations living in different regions is largely attributable to variation in ambient levels of trace elements in the local waters.

Certain elements may be concentrated in elements of the diet eg. strontium in herbage, zinc in meat, fluorine in tea, lead in Roman wine, and this may be an added source of excess or relative deficiency of the elements in the teeth. Again some health conditions may alter the ease with which an element is taken up. For example, when conditions in the mouth are acid, perhaps due to caries ore plaque-forming bacteria, fluorine and bromine levels become elevated regardless of the availability of these elements in the drinking water or food. Since one carious lesion will alter the chemistry of the whole mouth, analyses of a single tooth can only be interpreted in the context of the condition of the rest of the teeth.

| | Enamel % | Dentine % | Cementum % |
|---|---|---|---|
| Water | 2.3 | 13.2 | 32 |
| Organic | 1.7 | 17.5 | 22 |
| Ash | 96 | 69.3 | 46 |
| Ca | 36.1 | 35.3 | 35.5 |
| P | 17.3 | 17.1 | 17.1 |
| $CO_2$ | 3.0 | 4.0 | 4.4 |
| Mg | 0.5 | 1.2 | 0.9 |
| Na | 0.2 | 0.07 | 1.1 |
| K | 0.3 | 0.03 | 0.1 |
| Cl | 0.3 | 0.017 | 0.1 |
| F | 0.016 | 0.2 | 0.015 |
| S | 0.1 | | 0.6 |
| Cu | 0.01 | | |
| Si | 0.003 | | 0.04 |
| Fe | 0.0025 | | 0.09 |
| Zn | 0.16 | 0.018 | |
| Pb | - | 0.0071-0.037 | |

Also
Ce, La, Pr, Ne
Ag, Sr, Ba, Cd, Cr, Sn, Mn, Ti, Ni, V, Al, B,
Li, Se.

Table 1. Approximate composition of dental tissues. The basic mineral is hydroxy-apatite

$$\left[Ca_9{}^{++}(H_3O^+)_2(PO_4{}^{\equiv})_6(OH^-)_2\right] \quad \left[Ca^{++}\cdot Mg_{0.3}{}^{++}\cdot Na_{0.3}{}^{+}\cdot CO_3{}^{\equiv}\cdot cit_{0.3}{}^{\equiv}\right]$$

(Vernois this Volume). Dietary imbalance can also alter the proportions of some elements such as cadmium, the intake of which can be doubled if the diet is low in protein, calcium or iron (Hallenbeck 1984).

A final and most important source of trace elements , when considering archaeological or prehistoric material is through uptake from ground waters bathing the sediment in which the material has been buried. This post-mortem uptake of trace elements is an important source of error and confusion in the analysis and interpretation of dental material. It should be assumed that some degree of diagenesis has affected the chemistry of all buried material. One of the problems of historical anthropology is to elucidate the nature and extent of post-mortem diagenesis.

## Post-mortem uptake of trace elements

The uptake of trace elements by buried material, teeth or bone, depends on the availability of the elements in the ground waters, temperature, redox and acidity, as well as on the intrinsic chemistry and structure of the material. An intact tooth with its crown of dense, compact enamel and roots protected by dense, usually acellular cementum is relatively imperviuos to the movement of elements into the structure from the exterior surface. However the uptake of elements can and does take place readily from the surface of the pulp cavity into the dentine.

Studies of fluorine distribution in the dentine of a series of Romano-British teeth from Poundbury, Dorset that had been buried in calcareous soil in a low fluorine area for about 1500 years were carried out using PIXE techniques described by Coote and Sparks (1981) to scan the sectioned surface. In all the teeth it is quite clear that the fluorine concentration is highest on the surface and least towards the enamel-dentine boundary. This boundary acts as an effective barrier to trace element transport (Figure 1). The transport of fluorine appears to be on the surface of apatite crystals that are the essential component of the mineral phase of dentine. Later it is incorporated by ionic exchange with hydroxyl into the apatite lattice. Fluorapatite is more stable than hydroxyapatite, the main apatite form in hard tissues and the fluorine usually remains fixed in the mineral.

The concentration gradients seen in figures 1 and 2 are a function of time and Coote and Nelson (1984) have suggested that the slope of the gradient can be used to calculate duration of burial and therefore the age of the specimen.

Where the surface enamel or cementum is damaged, fluorine is readily taken up from the outer surface (Figure 2). The extent of post-mortem uptake can be seen easily in scans across such damaged areas and it is by this means that a possible source of error in failing to recognise post-mortem exchange can be avoided. For it is not always possible to distinguish life-time gradients from gradients developed in the ground if only the pulp to dentino-enamel boundary gradients are considered. Ancient material and material buried in sediments bathed by fluorine rich ground-waters, usually have levels of fluorine that would cause dental defects of tooth

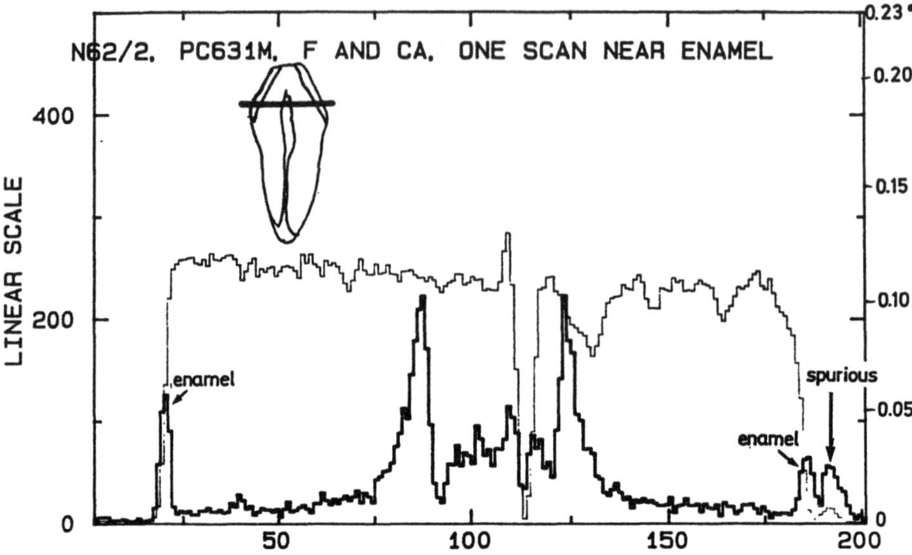


Fig.1. PIXE scan showing variation in concentration of fluorine (black line) and calcium (grey line) across the intact, undamadged crown of a tooth from a 5th Century AD burial. The greatest concentration of fluorine is in the dentine adjacent to the pulp cavity. Scans prepared by G. Coote CSIR New Zealand.

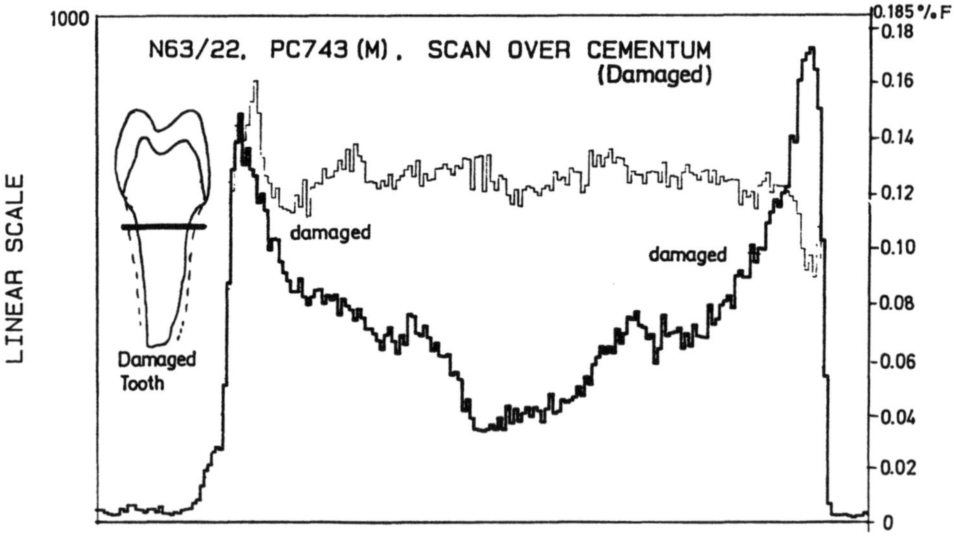


Fig. 2. PIXE scan showing peripheral uptake of fluorine into the dentine of the root of a tooth in which the cementum layer has been damaged. Dense acellular cementum normally inhibits the uptake of trace elements. Scans prepared by G.Çoote, CSIR New Zealand

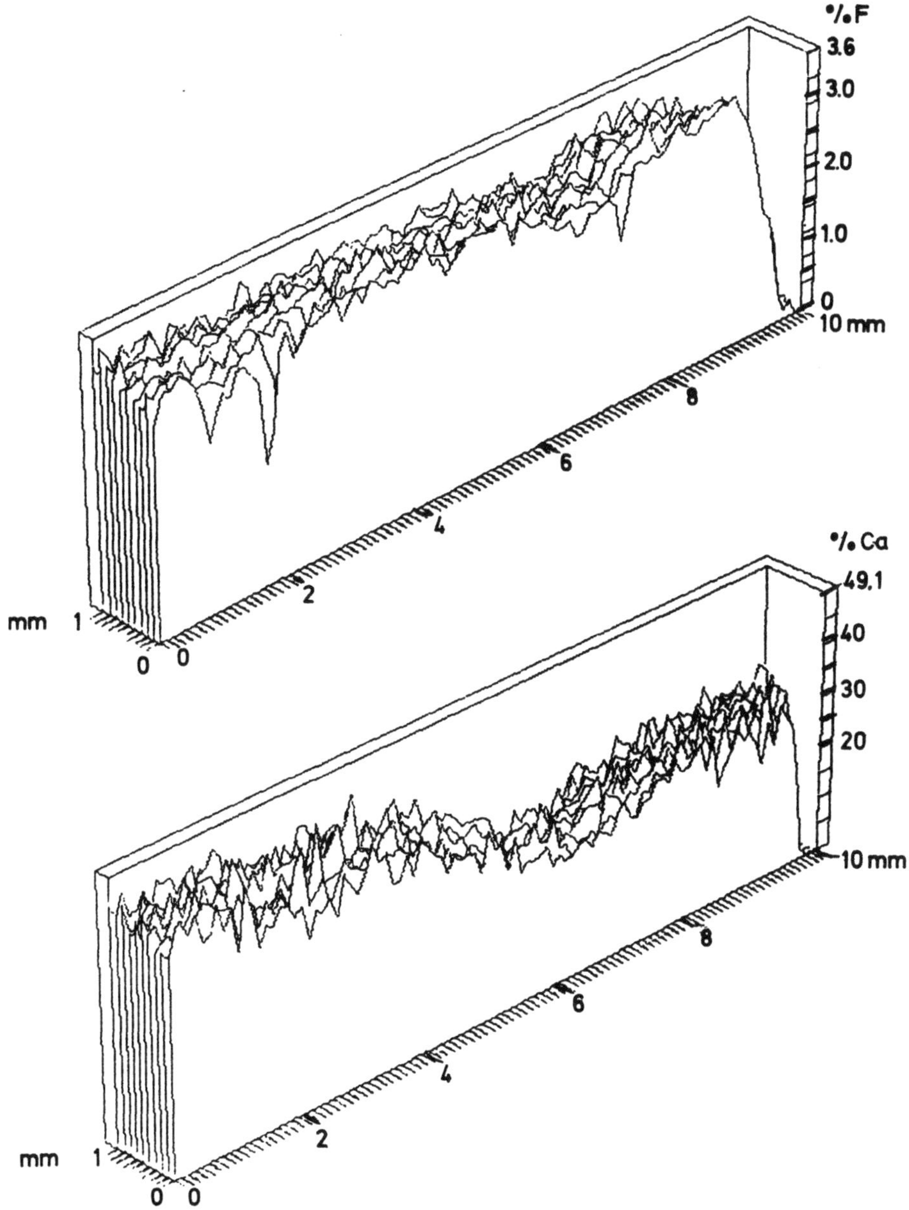


Figure 3. PIXE scans of fluorine (above) and calcium (below) concentrations in a fossil rhinoceros tooth from Olduvai Gorge. About 1.7 million years old. The fluorine distribution has reached a plateau and the concentration of 3.6% implies complete replacement of the $OH^-$ by $F^-$ in the apatite crystals. Scans prepared by G. Coote, CSIR New Zealand.

fluorosis in the living and/or profiles with shallow gradients that have developed over a period longer than a lifetime (Figure 3).

While the shape of the distribution profile in the dentine for a given element is a function of accumulation time, the concentration is largely dependent on ambient levels in the environment and in certain cases (cerium, uranium) on the valency state (see Williams and Marlow 1987; Williams this volume). Concentrations of over 3% fluorine recorded in fossils from East Africa (Figure 3) are exceptional. However, concentrations may build up over a period of time so that fossils from older strata tend to contain higher levels of such elements as fluorine, uranium or manganese than do fossils from younger strata at the same time. Comparison of concentrations of trace elements in fossil bone and teeth from a site is the basis of the relative dating technique for identifying intrusive or derived material that was put forward by Kenneth Oakley (1964).

Uranium is widespread at very low levels in groundwaters and is extremely readily taken up by buried hard tissues. It can usually be detected in a tooth that has been exposed to groundwaters for any length of time. The levels in a molar tooth of a fossil elephant supposedly recovered from the gravels at Piltdown were sufficiently high for an autoradiograph of the distribution to be possible (Weiner et al. 1955) and incidently its provenance to be suspect since such high uranium levels are not found in young deposits in the United Kingdom. Significant levels of uranium are not found in the teeth of the living and the detection of this element in excavated material is a sure indication of post-mortem contamination.

## Trace element distribution in the teeth of the living

Apart from post-mortem alteration, one of the biggest complications to the study of life-time levels of trace elements in teeth is that the concentration of almost any element concerned varies according to position in the tooth and age of the individual.

Detailed studies of sections of enamel by Robinson et al. (1986) have shown that fluorine is concentrated in the enamel surface. This developmental surface concentration is only 1/20 mm thick (Brudevold and Söremark 1967) and is easily removed by attrition. During an individual's lifetime fluorine continues to be absorbed by the enamel, especially locally in more acid areas around the gum margin and in the vicinity of dental plaque. Secondary uptake is greatest by enamel that is affected by a carious lesion. Fluorine concentration continues to increase throughout the life of the individual. A steep gradient from outside to enamel-dentine junction is, however, maintained (Figure 4).

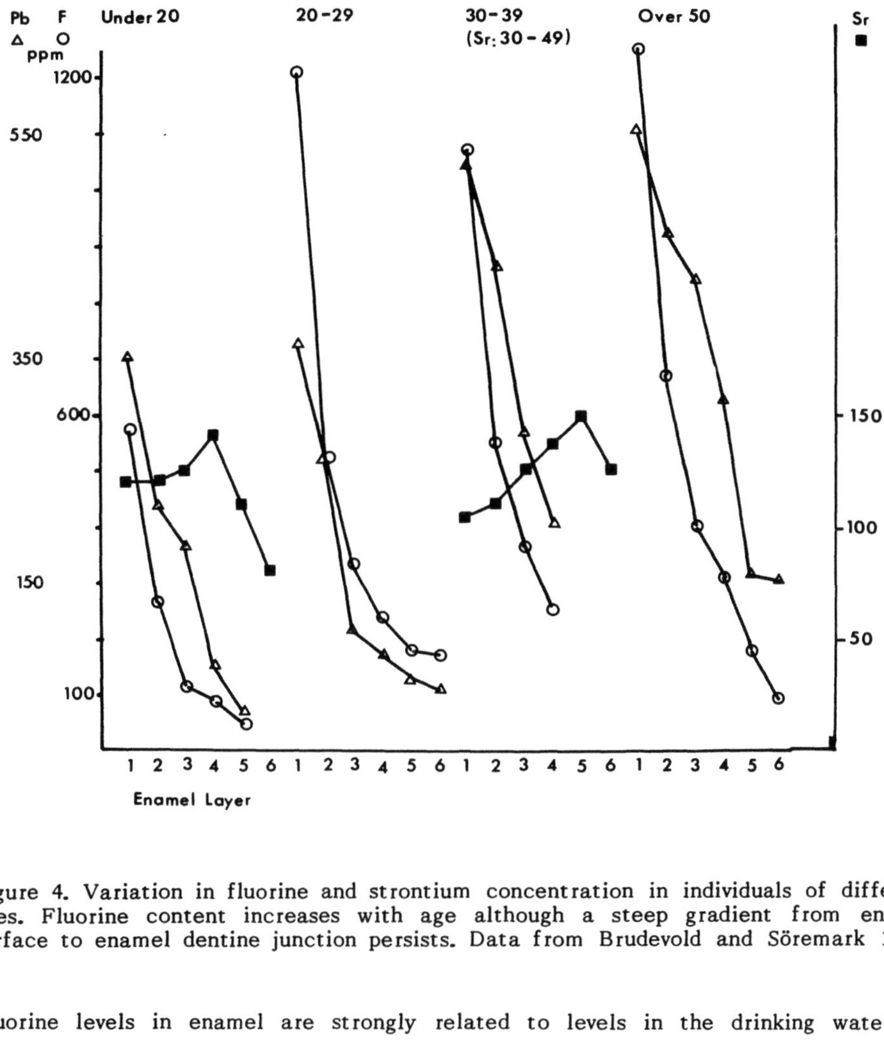


Figure 4. Variation in fluorine and strontium concentration in individuals of different ages. Fluorine content increases with age although a steep gradient from enamel surface to enamel dentine junction persists. Data from Brudevold and Söremark 1967.

Fluorine levels in enamel are strongly related to levels in the drinking water. In areas of endemic fluorosis these may reach 9000 ppm. In these circumstances the teeth become marked with brown spots and manifest enamel defects.

The distribution of enamel of other trace elements very often follows the same pattern as fluorine. Zinc, lead, cadmium and tin all are concentrated in the surface of the enamel. Strontium and copper in contrast are fairly evenly or randomly distributed in the enamel. Nor does the concentration of these elements appear to increase with age (Brudevold and Söremark 1967) (Figures 4, 5 and 6). Magnesium apparently has a reversed gradient with highest concentrations at the enamel-dentine boundary (Robinson et al. 1986). The concentration gradients are developed in unerupted teeth (Figure 6) and are maintained thoughout the life of the tooth.

Post-mortem uptake, if any, is imposed on the natural life-time gradient. It follows that the existence of a gradient does not necessarily imply post-mortem contamination.

The distribution of trace elements in dentine characteristically follows a gradient of highest concentration at the pulp surface falling to a minimum near the dentine enamel boundary. The gradients are maintained throughout life while concentration increases as it does in enamel (Trautz 1967). Dentine contains between two and three times the concentrations of enamel (Jenkins and Speirs 1954).

Deciduous teeth manifest the same concentration gradients for fluorine as do permanent teeth but the concentrations themselves are much lower with 600-700 ppm at the surface in low fluorine areas and about 1000ppm in an area with 1ppm fluorine (Iijima and Katayama 1985). The reason for this may be, in part because the deciduous teeth develop in a shorter time than do the permanent teeth and are therefore less able to acquire fluorine and because part of their development is in utero or while the child is suckling and fluorine does not readily cross the placenta or mammary glands (Underwood 1971).

In contrast Whittaker and Stack (1984) found much higher levels of lead and zinc in erupted deciduous teeth than in permanent teeth that had been buried in the ground. The levels were lower when the teeth were first abrasively cleaned to remove the outer surface and the question arises as to whether the higher surface levels had been acquired post-mortem by transfer from ground waters or in vivo by ectopic uptake. A number of trace elements including fluorine, zinc, lead, radium, uranium and chlorine build up after eruption because of reactions with oral fluids (Brudevold and Söremark 1967).

## Trace element accumulation

The chemistry of elemental uptake is governed by the physico-chemical properties of the apatite crystals and by the water in the intercrystalline spaces which serves as a medium for diffusion. Due to restricted diffusion chemical activity is limited to the outer portion of the enamel. The relatively large size and close packing of the enamel crystals means that enamel has a greater stability than dentine or bone. The reactive capacity of the mineral depends on both the surface area of the crystals and the volume, content and character of the spaces between. The changes that occur involve mainly the surface of enamel crystals and the inner part of the lattice is not affected.

Fluorine is taken up by direct contact with water. It diffuses quickly into the hydration layer, is fixed in the surface of the hydroxyapatite by replacing hydroxyl or bicarbonate and there is almost no migration to the body of the crystal. Even in high fluorine areas the concentration is far below the saturation of pure fluorapatite (3.6%). The more fluorine saturated the surface enamel is at the time of eruption, the less tendency it has to accumulate additional amounts (Brudevold and Söremark 1967).

Fluorine uptake appears to be increased in high altitudes. The effect of low atmospheric pressure on fluorine metabolism has been tested. This may be a major factor in explaining the greater severity of fluorosis in Kenya compared with Europe (Manji, Baelam and Fejerskov 1986). Ainsworth (1933) was the first to report that mottled teeth showed a marked protection from caries and two ppm in fluorine in the water will cause mottling of the dental enamel. The density of the enamel is correlated with the fluorine level and this may influence caries protection.

Today, chronic endemic fluorosis results primarily from consumption of highly fluoridated waters which are derived from deep wells or bores in which fluorine comes from deep-seated rock formations rather than from surface leaching. Surface water supplies - the normal source of water supply in the past - are not usually unduly high in fluorine.

Lead content in enamel reflects body lead burdens. Although lead will cross the placenta, pre-eruptive deposition is small and there is a relationship between ingestion of lead and deposition in dental tissue. Thus lead levels increase with age in teeth (Figures 5,6). That metals are labile in teeth is concluded from the work of Jenkins (1979) who demonstrated that lead levels were halved from the age of 5-12 years. Children absorb lead more efficiently than adults; their deciduous teeth may also be more permeable to lead in the environment of the mouth. Consequently lead levels in deciduous teeth are usually higher than in permanent teeth (Needleman et al. 1972). The resorbing apices of deciduous teeth may contain higher lead levels than the remaining dentine (Stack unpublished observation). The distribution of lead in dentine is similar to that of fluorine with a gradient from the pulp (Trautz 1967). The distribution in enamel is also similar to that of fluorine. The lead may exchange with the calcium in the apatite by first being absorbed on the crystal surface.

Some attempts have been made to correlate the concentrations of lead in different tooth structures and it has been concluded that the dentine concentration is very approximately equal to the whole tooth concentration (Al-Naimi, Edmonds and Fremlin (1980).

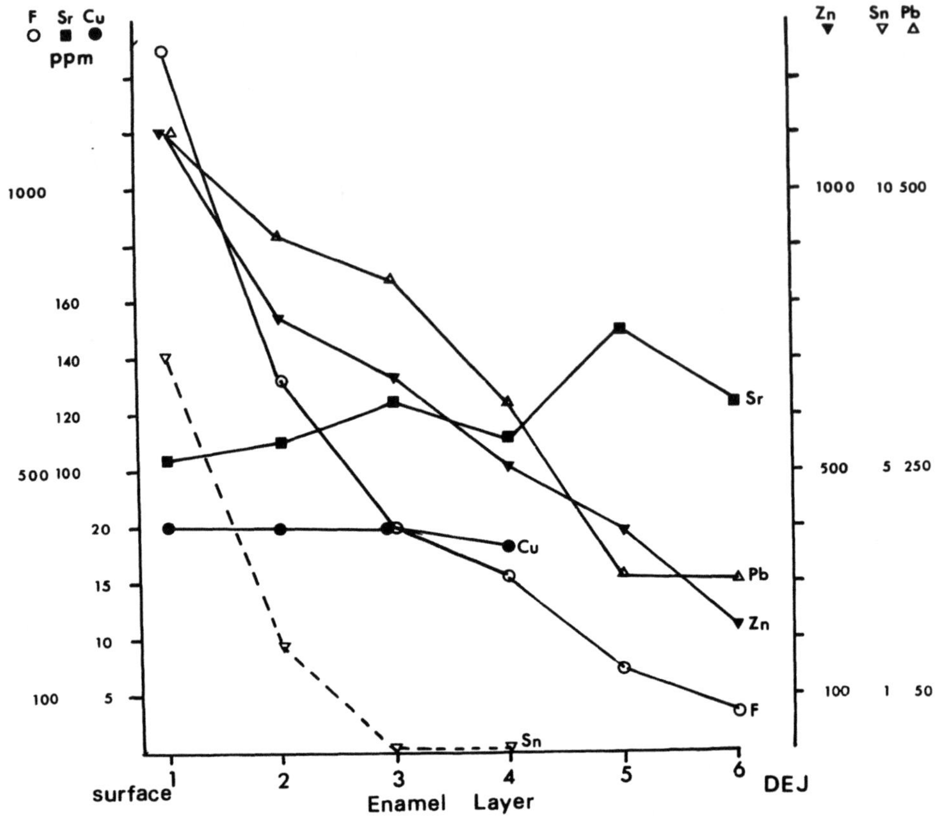


Fig. 5. Concentration profiles of trace elements in the enamel of erupted teeth of individuals aged over 30 years. Data from Brudevold and Söremark 1967.

Steenhout (1982) has shown that the first order kinetic governing the transfer of lead from blood to teeth is about 1.85 $yr^{-1}$. Lead elimination can be neglected, as dental tissues once fully formed do not undergo the remodelling processes as do bones. Where lead exposure in a given geographic area is constant, lead concentration in teeth is a linear function of time. The slope of the regression line is directly related to the mean blood-lead level. There is no difference between the sexes in the accumulation of lead in teeth (unless the environment of exposure differs).

Zinc is the only other trace element which has been found to occur in human enamel in concentrations approaching those of fluorine. It also accumulates in surface enamel. The high concentrations present in the enamel of unerupted teeth show that most of the zinc is deposited before eruption and although erupted teeth show greater concentrations than unerupted teeth from the same geographical area, a gradual increase in concentration with age has not been found (Figures 5,6) (Brudevold and Söremark 1967). Overall levels of zinc are, like lead, higher in deciduous teeth than in permanent teeth (Whittaker and Stack 1984).

Strontium is present in enamel and dentine and is incorporated in the apatite crystal by absorption and substitution. The concentration varies in different teeth but in general the levels correlate with concentrations in the food and especially the drinking-water (Brudevold and Söremark 1967). The main uptake occurs during formation of the teeth so that there is little variation within the tooth or with age although there may be some further deposition in secondary dentine or cementum (Trautz 1967). It follows that strontium levels in a tooth probably correlate with the dietary levels at the time of maturation of the tooth.

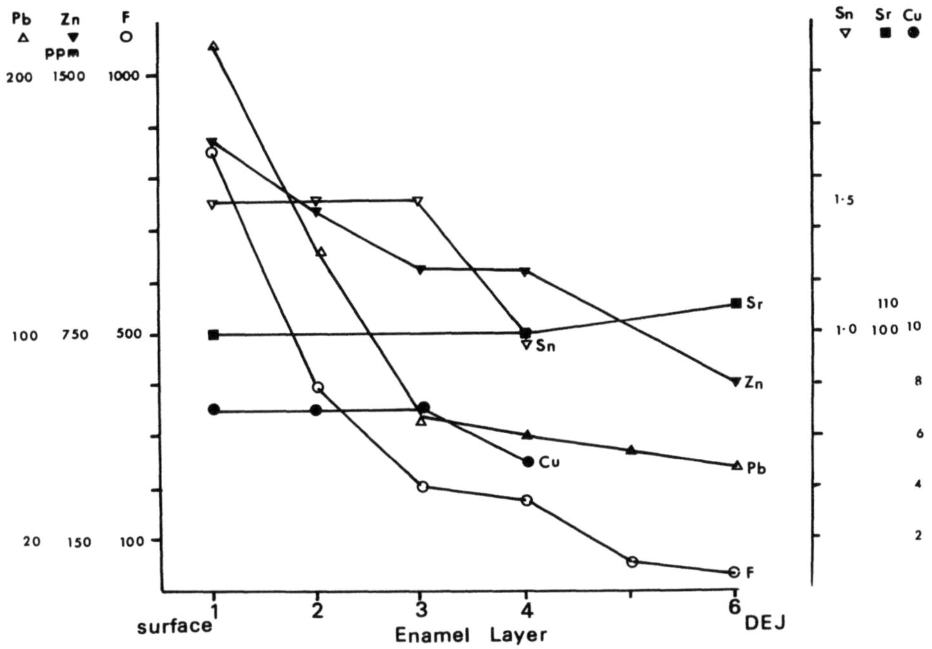


Fig. 6. Concentration profiles of trace elements in the enamel of unerupted teeth. Data from Brudevold and Söremark 1967.

Cadmium occurs in nature with zinc, lead and copper and some exposure to cadmium must result from the use of these metals. The placenta is an effective barrier and cadmium levels in the blood of the new-born are 50% less than in the maternal blood. A diet low in calcium, iron or protein can double the intake as can smoking. Wheat and rice grown on polluted soil can accumulate high amounts and, since it is stored in the kidneys and the liver, these organs in mammals can contain high amounts (Hallenbeck 1984, Bernard and Lauwerys 1984).

Molybdenum may lower the prevalence of caries while small amounts of selenium in the diet during development of the teeth increases the incidence of caries. This is true both of low and high fluorine areas (Underwood 1971).

Iron accumulates in the outer enamel as does tin, while magnesium increases towards the enamel-dentine junction: copper shows no discernable distribution pattern. Neither copper nor magnesium concentrations change with age. Copper and selenium are found predominantly in protein. It is possible for variation in copper concentrations between individuals to reflect only the differences in protein preservation. It follows that analysis of enamel which contains negligible amounts of protein avoids this problem (Bumstead 1985).

## Interpretation of trace element variation

It can be seen that trace element levels in the human skeleton can rarely be taken at face value, nor indeed can caries frequencies or pathology. While varying levels can reveal much about the dating, age, health or diet of individuals in a population as well as throwing light on environmental conditions of burial, cultural practices, diet and food adulteration in the past, great care is needed in the selection of samples and in the interpretation of results. It is doubtful if single element studies or analysis of a single individual are of any value.

Since that given the correct conditions, almost any groundwater borne element can probably be taken up by dental tissues, it is important to identify this uptake before the original composition of excavated material can be described. This can to some extent be achieved by comparing trace element composition of the burial medium with that of the excavated teeth or bone (Edward et al. 1984), but Henderson et al. (1983) and Williams (this volume) have shown clearly that the affinity of some elements, notably the rare earths (cerium) and uranium for apatite results in the accumulation of these elements in the biological material and their depletion in the immediately surrounding sediment. Elemental uptake appears to be greatest soon after burial and probably slows thereafter as equilibrium with the groundwaters is attained. In dry compacted and indurated matrices elemental transfer must be minimal.

Enamel of all the biological tissues is the least susceptible to alteration. It is by far the hardest, most dense and most highly mineralised of the mammalian tissues and because of this has the greatest potential to provide historical data about the period of tissue formation and its subsequent functional period in the oral cavity (Robinson et al. 1986), but maturing enamel of unerupted and recently erupted teeth is susceptible to post-mortem alterations related to the burial environment due to its porous nature.

Since trace element levels in vivo are usually primarily dependent on levels in the drinking water and only secondly related to levels in food, a comparison with modern material from the same animal, or one with the same water intake, is another check on post-mortem alteration (Edward et al. 1984; Francalacci and Borgognini Tarli this volume).

Single element studies can be misleading since burial contamination cannot easily be identified (Patterson et al. 1987). If the series sampled is large enough, however, the frequency distribution of the logarithms of the values will have a Gaussian distribution with marked negative skewing and only a few individuals with greatly elevated levels (Drasch et al. 1987). By this means it can be seen that many of the lead data for Romano-British bones are invalid as indicating life time intake. For example the lead values for bones from Henley Wood do not represent life time levels while those from Poundbı Camp probably do (Waldron et al. 1976).

Perhaps the most reliable method of identifying trace element differences in teeth, whether due to original intake or post-mortem accumulation, is by the use of ratios of antagonistic elements (e.g. Ba/Ca, Pb/Ca, Sr/Ca) (Patterson et al. 1987), or of complementary pairs of elements (e.g. Sr/Zn) (Edward et al. 1984).

There are now many studies of excavated material that demonstrate differences in trace element levels between sub-groups or at different time levels. These must reflect changes in dietary intake. Whereas studies on bone will reflect life time accumulation, those on teeth will primarily indicate levels at the time of maturation of the tooth. Final mineralisation of a tooth proceeds from crown tip to neck cutting across the initial incremental growth. In diet changes during the maturation process this could be reflected in differences between parts of the tooth and between teeth that develop at different rates or at different times.

Strontium and zinc are mainly deposited during the formation of the tooth and can therefore be used to monitor changes in dietary intake. Stack (1986) has inferred from Sr/Zn ratios in selected teeth that some ancient Nubians had a vegetarian diet. He also found iron levels to be so low as to suggest iron deficiency during tooth formation. Grupe (1986) has exploited the fact that zinc levels may be depleted in weanling children to determine the age of weaning.

Since natural fluorine has such an influence on caries rate, it should be taken into account when considering caries rates in past populations. Dental caries becomes a feature of human populations in Britain from Roman times. The real implications of this phenomenon are complicated by the fact that many of the data derive from populations which lived in low fluorine areas and by the fact that exposure to lead and cadmium in the diet increase susceptibility to caries development. The Romans it is well known, by their cooking methods and the use of boiled down fruit juices (sapa, defrutum) containing lead, were exposed to high concentrations of dietary lead.

At Poundbury Camp, Dorset, elevated levels of lead have been found in the bones (Waldron et al. 1976) and teeth (Whittaker and Stack 1986) of many individuals. The caries frequency is greater than that found for other populations of the period in England. This may be a direct consequence of the high dietary lead or it may be an indication that refined flours were being used by a sophisticated and romanised population. It may also reflect the older age composition of the population since caries frequency increases with age. That high dietary lead is not only the factor influencing caries rate is indicated by the lack of correlation between mean lead levels and caries frequency in the different parts of the cemetery.

Acknowledgements

I should like to thank Dr Graeme Coote, New Zealand, for the preparation of the PIXE scans; Dr Terry Williams, Department of Mineralogy, BM(NH), UK for useful discussion. My grateful thanks too to Edna Richards for preparing the typescript.

References

Ainsworth NJ (1933) Mottled teeth. Br dent J 60: 233

Al-Naimi T, Edmunds MI & Fremlin JH (1980) The distribution of lead in human teeth, using changed particle activation analysis. Phys Med Biol 25: 719-726

Bernard A & Lauwerys R (1984) Cadmium in human population. Experientia 40: 143-152

Brudevold F & Söremark R (1967) Chemistry of the mineral phase of enamel. In: Miles AEW (ed) Structural and chemical organisation of teeth. Vol II. Academic Press New York London. pp 247-277

Bumstead MP (1985) Past human behaviour from bone chemical analysis - respects and prospects. J Hum Evol 14: 539-551

Coote G & Nelson P (1984) Diffusion profiles of fluorine in bones and teeth: their measurement and application. In: New developments in dating the past. 54th NAZAAS Congs Canberra

Coote GE & Sparks RJ (1981) Fluorine concentration profiles in archaeological bone: an application of a Nuclear Microprobe. New Zealand J Archaeol 3: 21-32

Drasch GA, Bohm J & Baur C (1987) Lead in human bones. Investigations on an occupationally non-exposed population in Southern Bavaria (FRG) I Adults. Science of the Total Environ 64: 303-315

Edward J, Fossey JM & Yaffe L (1984) Analysis by neutron activation of human bone from the Hellenistic cemetery at Asine, Greece. J Field Archaeol 11: 34-46

Francalacci P & Borgognini Tarli S this volume

Grupe G (1986) Rekonstruktion bevölkerungsbiologischer Parameter aus dem Elementgehalt bodengelagerter Knochen. In: Herrmann B (ed) Innovative Trends in Prehistoric Anthropology. Mitt Berl Ges Anthrop Ethnol Urgesch 7: 39-44

Hallenbeck WH (1984) Human health effects of exposure to cadmium. Experimentia 40: 136-142

Henderson P, Marlow CA, Molleson TI & Williams CT (1983) Patterns of chemical change during bone fossilisation. Nature Lond 306: 358-360

Iijima Y & Katayama T (1985) Fluoride concentration in deciduous enamel in high and low fluoride areas. Caries Res 19: 262

Jenkins GN & Speirs RL (1954) Distribution of fluorine in human enamel. J Physiol 121: 21-28

Jenkins WG (1979) Lead levels in deciduous teeth in the Caernarvonshire area. MScD Thesis U of Wales

Manji F, Baelam V & Fejerskov O (1986) Fluoride altitude and dental fluorosis. Caries Res 20: 473

Needleman HL, Tuncay OC & Shapiro LM (1972) Lead levels in deciduous teeth of urban and suburban American children. Nature Lond 235: 111-112

Oakley KP (1964) The problem of Man's antiquity: an historical survey. Bull Br Mus Nat Hist Geol Ser 9: pp155

Patterson CC, Shirahata H & Ericson JE (1987) Lead in ancient human bones and its relevance to historical developments of social problems with lead. Sci Total Envir 61: 167-200

Robinson C, Kirkham J, Weatherell JA & Strong M (1986) Dental enamel - a living fossil. In: Cruwys E & Foley RA (eds) Teeth and Anthropology. BAR Intern Ser 291: 31-54

Stack MV (1986) Trace elements in teeth of Egyptians and Nubians. In: Cruwys E & Fowley RA (eds) Teeth and Anthropology. BAR Intern Ser 291: 225-231

Steenhout A (1982) Kinetics of lead storage in teeth and bones: an epidemiologic approach. Archs Environ Health 37: 224-230

Trautz OR (1967) Crystalline organisation of dental mineral. In: Miles AEW (ed) Structural and chemical organisation of teeth, Vol II. Academic Press New York London pp 201-246

Underwood EJ (1971) Trace elements in human and animal nutrition, 3rd ed. Academic Press New York London, pp 543

Vernois V this volume

Waldron HA, Mackie A & Townshend A (1976) The lead content of some Romano-British bones. Archaeometry 18: 221-227

Weiner JS et al (1955) Further contributions to the solution of the Piltdown problem. Bull Br Mus Nat Hist Geol Ser 2: pp 282

Whittaker DK & Stack MV (1984) The lead, cadmium and zinc content of some Romano-British teeth. Archaeometry 26: 37-42

Williams CT this volume

Williams CT & Marlow CA (1987) Uranium and thorium distributions in fossil bones from Olduvai Gorge, Tanzania and Kanam, Kenya. J Archaeol Sci 14: 297-309

# CHEMICAL ANALYSIS OF HUMAN DENTAL ENAMEL FROM ARCHAEOLOGICAL SITES

V. Vernois*, M. Ung Bao*, N. Deschamps**
* Laboratoire d'Histologie
Faculté de Chirurgie Dentaire, Paris V
1, rue M. Arnoux
92120 Montrouge, France
** Laboratoire "Pierre Sue" C.N.R.S.
C.E.N./Saclay Bât 637
91191 Gif sur Yvette Cedex, France

The high degree of mineralization is the most important characteristic of the dental enamel. Therefore, the teeth are often well preserved during long burial periods, contrary to bone tissues at least in certain cases. During its embryologic formation, enamel developes from epithelial tissue. The growth of enamel is a result of the mineralization of an extracellular matrix secreted by ameloblastic cells. At the end of its formation, the ameloblasts degenerate and are eliminated. The mature enamel is thus constituted of an acellular structure.

Issued from an epithelial tissue, the dental enamel is comparable to the overlying tissues of the organisms (e.g. skin, intestinal mucous...), and appears very different from the other mineralized tissues like bone or dentine which are issued from connective tissues. This is the reason why the dental enamel plays a very important role of a biological interface.

Its position in the buccal cavity allows some various exchanges between the inner part of the organism (across the enamel-dentine junction) and the external surroundings of the individual (consecutive to the nutritional intakes). The metabolic processes of enamel (post-eruptive growth) are depending on the presence of salivary fluids. In relation to the nutritional intakes, the salivary components permit some chemical elements to be incorporated into the enamel matrix. A biological balance exists between enamel and saliva, which ensures a partial rebuilt and a kind of turn-over despite the lack of cells.

However, disturbances of this biological system in relation, for instance, to irregularities of the buccal physiology, are able to interrupt this balance. In such cases, enamel is affected by pathological conditions as carious disease or calculus deposition. After death of the individual, these processes are interrupted and no more biological exchange can take place.

In order to study the physio-pathological state of enamel removed from ancient individuals, it was necessary to evaluate whether the chemical constitution of the enamel matrix was able to restore its biological behaviour during life time.

## Material and methods

The human dental samples come from three ancient necropoles situated in the south of France (Vieil Toulouse; Venerque; Montlaur) and dated from the first to the ninth century AD. Some archaeological studies have been carried out on each necropole by the "Groupe d'Étude Anthropologique des Sociétés Rurales" (1).

The investigation is interesting especially since the populations were of a rural, sedentary origin and were buried in situ. The chemical analysis of the soils and the vegetation allows the establishment of chemical environmental profiles.

A series of enamel samples removed from permanent and deciduous teeth was taken from a contemporary population in order to control the stability of the enamel composition during a long period of interment. A total of 120 enamel samples was taken from healthy permanent and deciduous teeth. In certain cases, several teeth from one and the same individual were analysed.

The enamel samples were put into cadmium containers (10 mm thickness) and irradiated with a neutron flux of $10^{13} n \cdot cm^{-2} sec^{-1}$ for 12 hours. Under such experimental conditions, the only chemical separation necessary concerns sodium due to its high concentration in dental enamel. This separation was carried out by the method of mineral exchange at acid conditions (2).

As, Au, Br, Cu, Eu, La, Mn, Sb, Sr, U, W and Zn concentrations were determined. Thus, more than 1400 values could be obtained.

The geographic differences of the populations tested and the physio-pathological variations amongst individuals within each population itself do affect the mean results. Associated with large standard deviations, most of the mean values are not suitable for data interpretation. Therefore, the concentration values obtained from all individuals examined were compared using a statistical analysis based on the correlation matrix: the multivariate analysis (3). This method permits the comparison of all elements throughout the entire sample.

By multielementary analysis, a correlation diagram is obtained which permits the observation of a simultaneous representation of two corresponding groups on the same map: the enamel samples and the chemical elements analysed. The axises are determined according to the specific dimensions of the statistical cloud. The coordinates of each point (samples and elements) represent statistical values only and not concentration values. The choice of axises ("1-2" or "2-3") is influenced by the position of the statistical cloud. The positions of the enamel samples associated with the position of the chemical elements determine the biological behaviour of each element in relation to the physio-pathological features of each sample.

Using the disjunctive tables, the mean values of each element calculated from about 200 archaeological samples serve as references. When the concentration of one element is lower than the mean value, it is assigned as value 0 and when its concentration exceeds the mean value, it is assigned as value 1.

## Results and discussion

Trace element analysis of enamel removed from contemporary teeth enabled us to define the different constitutive groups of elements present in the enamel matrix:

- copper, strontium and zinc concentrations represent a scale of values with a small dispersal of the standard deviation. These elements can be considered as representative for the chemical constitution of the mineral matrix (4).
- bromine and manganese concentrations often show large variations which can always be associated with a physio-pathological characteristic of the samples (5).
- the presence of arsenic, europium, lanthanum and uranium in dental enamel is particularly associated with the chemical environmental profile of the individuals (6). However, in spite of soil constitutions which can affect the enamel composition during interment, the presence of such elements in plants or in water sources may as well enter the living organism by nutritional intake. Further studies have shown that the incorporation of arsenic and rare earth elements into enamel depends on the presence of the biological systems (7,8); on the other hand, As, Eu, La and U concentrations were determined at the same levels in some enamel samples removed from freshly extracted modern teeth (9).

A comparison established between all ancient populations and a modern population does not reveal any significant difference concerning the origin of the samples (fig. 1).

FIGURE 1

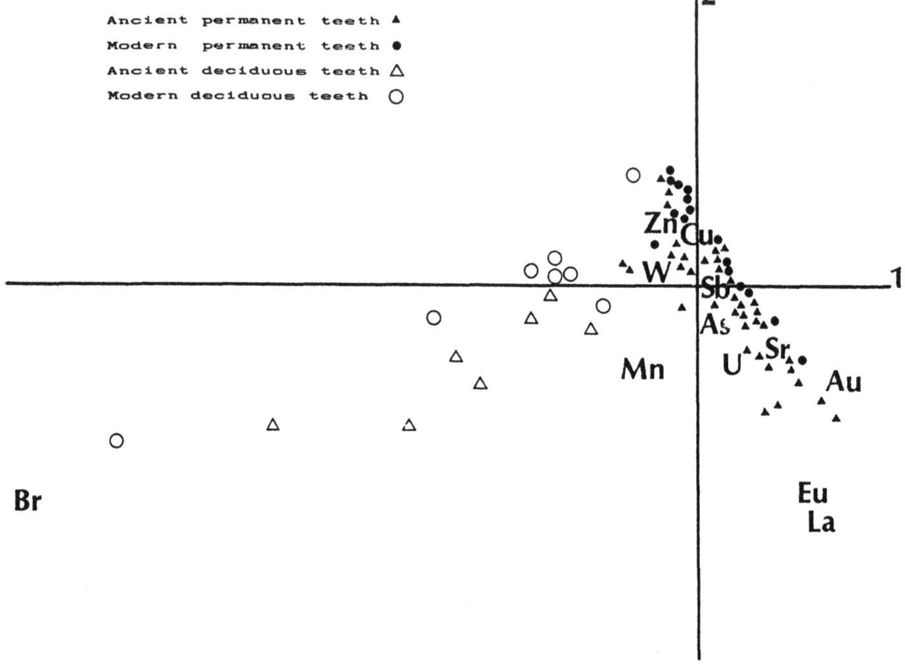


The main characteristic of the distribution of the samples is related to the specific bromine concentration of deciduous teeth (10). Therefore, two main groups can be determined on this diagram:

- a first group constituted of permanent teeth (established around the Sr and Zn positions)
- a second group constituted of deciduous teeth (associated with the excentric bromine position).

In each group, modern and ancient teeth are represented together.

This result shows that the chemical constitution of dental enamel is not affected by decomposition.

The investigation of the population from Montlaur shows a comparable distribution of enamel samples in relation to their physiological characteristics (fig. 2).

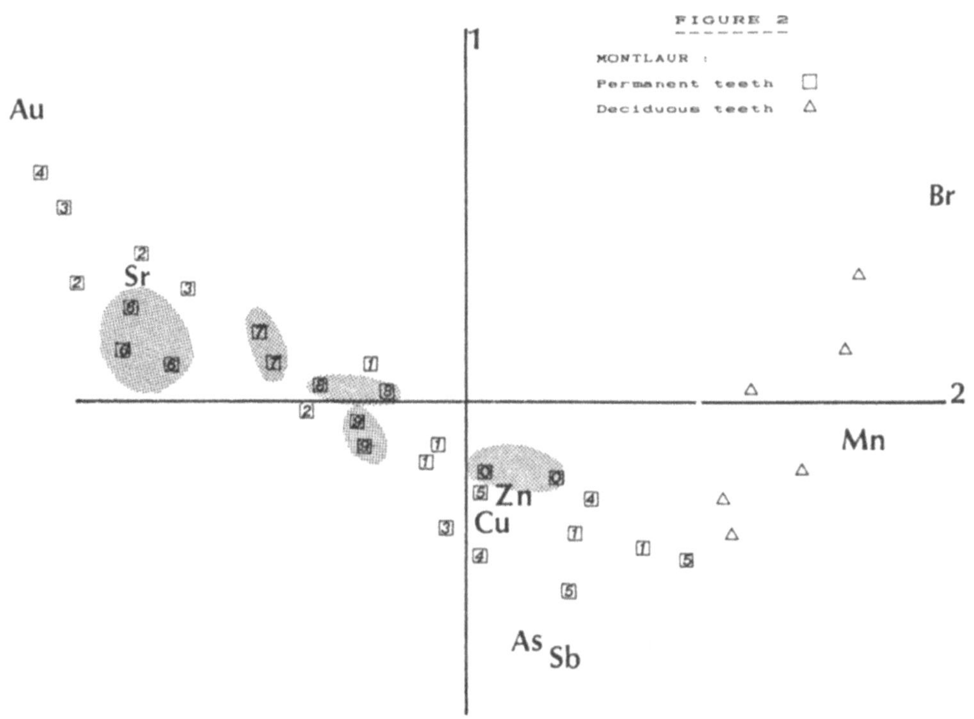


In most cases, the chemical composition of dental enamel seems to be specific for one individual: different samples removed from different permanent teeth of the same individual are grouped on the diagram because most of the concentration values present a small dispersal of the standard deviation (individuals number 6, 7, 8, 9 and 0). However, for certain individuals (number 1, 2, 3, 4 and 5), the enamel samples are scattered along the Zn-Sr axis in relation to a modification of the concentration levels of these elements.

Furthermore, when the disjunctive tables are used in order to compare the different ancient populations, specific modifications in Zn and Sr concentrations of these samples are observed (fig. 3).

Indeed, from the disjunctive tables, according to the "1-2" axises, Sr and Zn are the main discriminating factors. Thus, for these certain individuals, the samples present a particular situation contrary to the other samples which are grouped around the origin of the axises. The reason may certainly be due to the state of the buccal system of each individual.

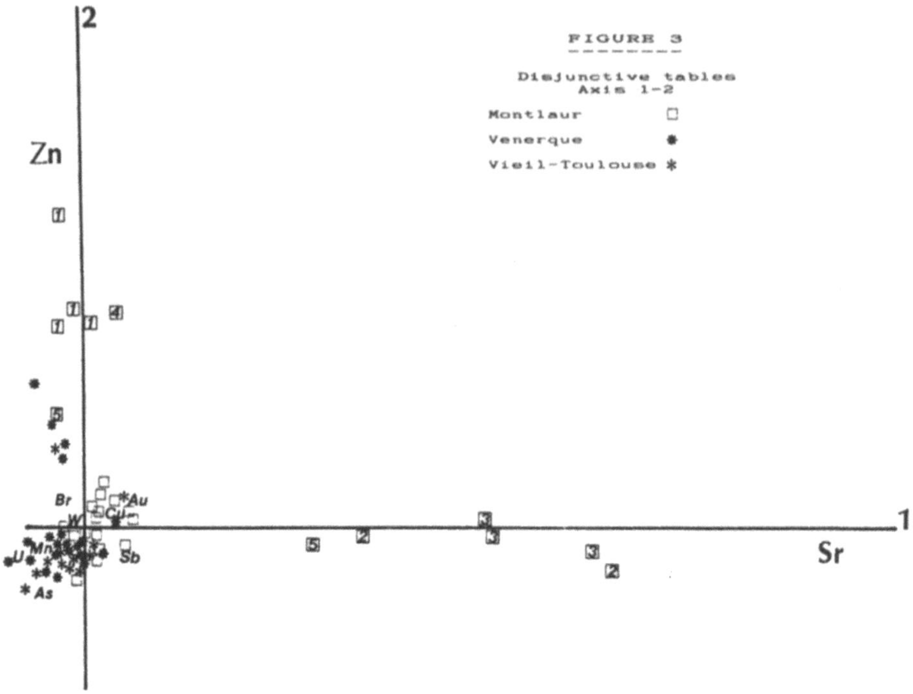


In spite of the clinically healthy nature of the samples, the chemical modifications of enamel are related to specific physio-pathological behaviour:

- Different teeth of different developmental steps were removed from one individual (number 1): presence of erupted and unerupted teeth.
- Disturbances in the salivary metabolism (presence of dental calculus) in relation to an abnormal deglutition (the forward position of the incisor group is characteristic of a lingual pressure) were observed in one individual (number 2).
- The chemical composition of teeth removed from one individual (number 3) which presented a carious disease (most of the other teeth were decayed) was affected early by the carious process, revealed by the lack of structural modification.

According to "2-3" axises, zinc and arsenic are the main discriminating factors (figure 4).

In this case, the physio-pathological differences between the samples do not appear any more. However, the particular high levels of arsenic in enamel removed from the population of Vieil-Toulouse permit us to distinguish between individuals from the various populations tested.

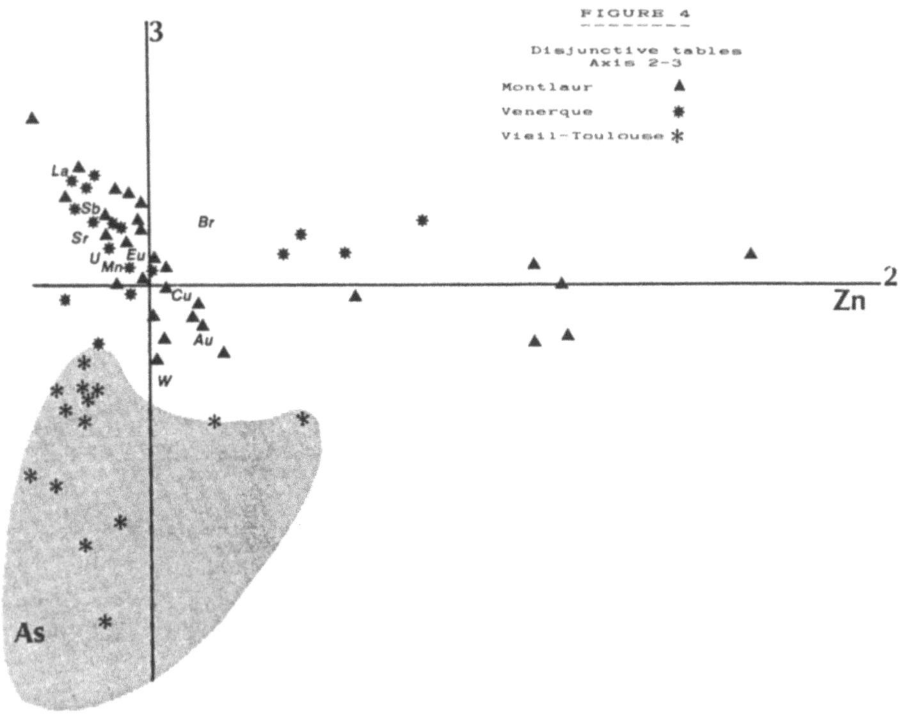


The chemical constitution of enamel is representative for the individual and its conditions of life, respectively. In order to study the relation between trace elements constitutive of dental enamel from burial sites and their biological behaviour during the individual's lifetime, the specific dental characters, the balance of the whole buccal system and the composition of the surroundings have to be regarded.

Given these conditions, dental enamel is considered a witness of the individual.

References

(1) Guilaine J, Duray H, Crubezy E, Vives E, Zammit J, Galabert J, Larrouy G, Arnaud J, Revel G, Deschamps N, Vernois V, Samadi A, Auffray A (15.11.1982) Anthropologie de populations médiévales en Languedoc. Compte-Rendu de fin d'étude d'une recherche financée par la Délégation Générale à la Recherche Scientifique et Technique. Action Concertée: Archéologie Métropolitaine. Décision d'Aide no 79.70398.

(2) Deschamps N, Vernois V, Calza F (1982) Dosage des oligoéléments de l'email dentaire par irradiation neutronique. J Radioanalyt Chem 70: 109-116.

(3) Benzecri (1980) L'analyse des données II. L'analyse des correspondances. Dunod Ed., Paris.

(4) Vernois V (1984) Les oligoéléments de l'email dentaire. Mémoire chimique de l'histologie biologique de l'individu. Thèse 3ème cycle Sciences Odontologiques, Paris V.

(5) Vernois V, Deschamps N, Ung Bao M, Vernois J (in press) Chemical analysis of dental enamel. Biological interest of trace elements. Microprobe and microtrace analysis.

(6) Davies BE, Anderson RJ (1987) The epidemiology of dental caries in relation to environmental trace elements. Experimentia 43, Birkhäuser Verlag.

(7) Remy-Ladam A(1987) Structures dentaires du rat et métabolisme de l'arsenic. Thèse de 2ème cycle Chirurgie Dentaire, Paris V.

(8) Vernois V, Deschamps N, Revel G, Septier D, Goldberg M (1982) Utilisation du dysporium en tant que traceur activable dans l'étude des structures dentaires. J Biol Buccale 10: 134-145.

(9) Ung Bao M (1986) Nouvelle méthode d'etude de la phase matricielle de l'email dentaire. Thèse de 2ème cycle Chirurgie Dentaire, Paris V.

(10) Vernois V, Deschamps N, Boulay A, Goldberg M (1984) Comparative study of the chemical behaviour of trace elements in dental enamel. In: Priest ND (ed) Metals in bone. Proceedings of EULEP Symposium. MTP Press. pp 407-415.

# TRACE ELEMENT CONTENT IN PREHISTORIC CREMATED HUMAN REMAINS

Bernd Herrmann and Gisela Grupe
Institut für Anthropologie der Georg August-Universität
Bürgerstr. 50
34oo Göttingen, FRG

Cremated human remains are an important source for the reconstruction of man-environment interactions, since cremation has been practised mainly or even exclusively during prehistoric periods. At least in central Europe some 60 generations cremated their death during Bronce and Iron Age and following periods, leaving numerous cremations, which tend to preserve better than not cremated bones (for references see Lange et al. 1987).

Incineration temperatures certainly have reached some 1000 °C, inducing extensive changes in the structural organization of the mineral matrix as well as the complete combustion of organic components. Trace element content of cremated bones might be influenced by thermal effects, thus causing concentrations patterns which may not be indicative for reconstructing environmental features. Therefore it seems reasonable to assume that at least concentrations of those trace elements also incorporated in the organic components should be influenced by heat exposure.

So far to our knowledge there is only one paper including data for cremations, i.e. two samples from Wisconsin/USA, suggesting no significant differences due to incineration (Price & Kavanagh 1982). No useful information regarding our question can be drawn from manuals on trace element contents in ashed bone (e.g. Iyengar et al. 1978), which are usually ashed at 600°C.

This paper is therefore done as a pilot study to test trace element analysis in cremated remains.

## Materials and method

Testing both intraindividual variability as well as the influence of location of the sample in the skeleton, five samples were taken from a cremation from Andernach

(FRG), dating to 2nd century A.D. The cremation was found in a glass amphora and belonged to a male approximately 50 years of age. Samples were taken from skull vault, rib, vertebra (corpus), innominate bone (ilium) and femoral shaft.

Interindividual variability was tested by analysing femoral shaft fragments from seven cremations coming from a Roman graveyard (1st century A.D.) from Urmitz (FRG). The cremations were sexed and aged as follows:

| sample # | sex | age (range) |
|---|---|---|
| 27 : 1 | probably female | 40 - 60 |
| 38 : 1 | probably male | 30 - 45 |
| 54 : 1 | male | 45 - 55 |
| 78 : 2 | male | 40 - 60 |
| 37 : 6 | male | 20 - 40 |
| 42 : 6 | male | 35 - 50 |
| 53 : 13 | non determinandum | 40 - 50 |

Data from these cremations were compared with femoral data from Andernach and one old adult female from Birkefehl (FRG) (# K 38), dating to 600 B.C.

Every sample appeared completely cremated with respect to it's macro- and micromorphological features. All cremations were choosen simply because of their availability at time of analysis.

Between 100 and 500 mg of each sample were solved in 25 ml $HNO_3$ ("suprapur" by Merck). The following elements were determined by ICP: Ca, P, Sr, Ba, Mg, Zn and Pb. Standard reference has been H-5 animal bone of the International Atomic Energy Agency.

The measurement error is less than 5% for Ca, P, Sr and Mg, less than 10% for Zn and Pb, and less than 20% for Ba. The analysis was carried out at the Hahn Meitner Institut, Berlin (West), by Dr. U. Rösick and staff, which is gratefully acknowledged.

## Results

The results are given in table 1 and table 2. As for statistic parameters given in table 2, they are only given as approximations. For comparison with recent data and those obtained from recent ashed bones cf. table 3.

TAB. 1: TRACE ELEMENT CONCENTRATIONS IN CREMATED SAMPLES

| range or error (%) | | | 5 | 5 | | 5 | 20 | 5 | 10 | 10 |
|---|---|---|---|---|---|---|---|---|---|---|
| sample | age | sex | Ca | P | Ca/P | Sr | Ba | Mg | Zn | Pb |
| | | | mg/g | mg/g | | ppm | ppm | ppm | ppm | ppm |
| Andernach: | 45-55 | m | | | | | | | | |
| skull | | | 316.8 | 142.8 | 2.22 | 147 | 8.6 | 2270 | 41 | 1.1 |
| rib | | | 388.9 | 172.2 | 2.26 | 124 | 12.4 | 2003 | 104 | 3.1 |
| vertebra | | | 235.3 | 103.3 | 2.28 | 76 | 5.5 | 1630 | 53 | 2.2 |
| hip | | | 468.8 | 212.6 | 2.21 | 159 | 8.9 | 3520 | 107 | 8.0 |
| femurshaft | | | 252.6 | 138.4 | 1.83 | 116 | 3.0 | 2250 | 58 | 0.5 |
| Urmitz (femurhsaft): | | | | | | | | | | |
| 27 : 1 | 40-60 | f | 358.4 | 166.6 | 2.15 | 150 | 10.0 | 2310 | 38 | 1.5 |
| 38 : 1 | 30-45 | m | 336.0 | 156.9 | 2.14 | 138 | 11.0 | 1590 | 30 | 0.9 |
| 54 : 1 | 45-55 | m | 359.7 | 166.5 | 2.16 | 202 | 7.2 | 1880 | 37 | 1.6 |
| 78 : 2 | 40-60 | m | 345.8 | 159.8 | 2.16 | 228 | 7.6 | 1550 | 58 | 2.1 |
| 58 : 13 | 40-50 | nd | 346.8 | 159.7 | 2.17 | 175 | 10.0 | 1490 | 36 | 1.6 |
| 37 : 6 | 20-40 | m | 349.6 | 162.6 | 2.15 | 171 | 6.7 | 1860 | 50 | 2.3 |
| 42 : 6 | 35-55 | m | 255.2 | 115.9 | 2.20 | 113 | 23.0 | 340 | 113 | 2.0 |
| Birkefehl (femurshaft): | | | | | | | | | | |
| K 38 | 40-60 | f | 388.7 | 180.4 | 2.15 | 371 | 11.0 | 720 | 20 | 0.9 |

**TAB. 2: MEANS, STANDARD DEVIATIONS AND COEFFICIENTS OF VARIATION OF TRACE ELEMENT CONCENTRATIONS**

| sample | Ca (mg/g) | P (mg/g) | Ca/P | Sr (ppm) | Ba (ppm) | Mg (ppm) | Zn (ppm) | Pb (ppm) |
|---|---|---|---|---|---|---|---|---|
| Andernach | | | | | | | | |
| (intraindividual variation, n=5) | | | | | | | | |
| mean | 332.5 | 153.9 | 2.16 | 124.4 | 7.7 | 2334.6 | 72.6 | 2.98 |
| standard deviation | 97.3 | 40.9 | 0,19 | 32.1 | 3.6 | 711.1 | 30.7 | 2.98 |
| coeff. var. | 0.29 | 0.27 | 0.09 | 0.26 | 0.47 | 0.30 | 0.42 | 1.00 |
| Urmitz | | | | | | | | |
| (interindividual variation, n=7) | | | | | | | | |
| mean | 349.4 | 162.0 | 2.16 | 177.3 | 8.8 | 1771.7 | 41.5 | 1.67 |
| standard deviation | 8.8 | 4.0 | 0.01 | 33.2 | 1.8 | 314.8 | 10.4 | 0.49 |
| coeff. var. | 0.03 | 0.02 | 0.005 | 0.19 | 0.21 | 0.18 | 0.25 | 0.29 |

TAB. 3: VARIABILITY OF TRACE ELEMENT CONTENTS

| sample | Ca | P | Sr | Ba | Mg | Zn | Pb |
|---|---|---|---|---|---|---|---|
| Andernach | 235 - | 103 - | 76 - | 3 - | 1630 - | 41 - | 0.5 - |
| (intraind.) | 469 | 173 | 159 | 13 | 3520 | 107 | 8.0 |
| Urmitz | 255 - | 116 - | 113 - | 6 - | 340 - | 30 - | 0.9 - |
| (interind. incl. 42 : 6) | 359 | 167 | 228 | 23 | 2310 | 113 | 2.3 |
| Urmitz 42 : 6 | 255 | 116 | 113 | 23 | 340 | 113 | 2.0 |
| Urmitz | 336 - | 156 - | 138 - | 6 - | 1490 - | 30 - | 0.9 - |
| (interind. excl. 42 : 6) | 359 | 167 | 228 | 11 | 2310 | 50 | 2.3 |
| Birkefehl | 389 | 180 | 371 | 11 | 720 | 20 | 0.9 |
| recent | 170 - | 103 - | 42 - | 4 - | 700 - | 50 - | 18 - |
| (dry-weight, fat free) | 274 | 174 | 237 | 20 | 980 | 170 | 70 |
| recent | 309 - | 155 | 90 - | 7 - | 5300 - | 190 | 20 - |
| (ash) | 390 | | 172 | 29 | 9370 | | 69 |

## Discussion

### 1. Basic assumptions

Surprisingly, the Ca/P ratios found (average 2.16) are very similar to that of native bone (tab. 1). An explanations could be that hydroxyapatite is transformed at temperatures above 800°C into high temperature modification (tricalcium-phosphates). These show strong re-reactions to hydroxyapatite under presence of $H_2O$ (Hummel & Schutkowski). After a few months only experimentally incinerated bones exhibit a typical apatite-like XRD-pattern, thus indicating that a cremation after a very short time will have rechanged into hydroxyapatite, exhibiting a "physiological" Ca/P ratio. With respect to this reversibility of reactions it is assumed that a cremation is exposed in principle to similar diagenetic processes as an unburnt bone, except for those factors which are based on the decomposition of the organic substance.

The concentration of Mg shows a significant increase in cremations (tab. 3). If these values are compared with both recent dry weight/fat free bones and recent ashed bones it is obvious that Mg enrichment is obtained from the organic matrix and concentrated by thermal shrinkage of bone. Since weight before incineration is influenced by organic compounds, which are completely burnt, their remaining ashes are added to a shrinked bone. So there are two overlapping processes causing an increase of content: shrinkage and mineral residues from the organic fraction.

This holds also true for Ca and P. At least Ca is higher concentrated in recend ashed bones (tab. 3) reaching values resembling those of the cremated samples. Phosphorous patterns are not that clear (tab. 3), as no ranges for phosphorous content are givenby Iyengar et al. (1978). However, it is assumed from the results that Ca and P are also "accumulated" in cremated bones by combustion of the organic fraction. Since Ca/P ratios in cremations are very close to those of uncremated bones, it is understood why Price and Kavanagh (1982) speculated from their two samples that "cremation did not introduce significant difference". Obviously this statement can not be confirmed by our study.

Sr and Ba, which are physiologically concentrated in the bone mineral, were not enriched through incineration processes as is shown by the values. Although Pb is a mineral bound element too, it's concentration is very low. At least Pb is lost under cremation at temperatures to be assumed (Hedrich).

Finally, tab. 3 shows rather low values for Zn. Zn values are strongly depending on availability of dairy products and meat. Low values could be explained by different access to those food resources, as seems to be supported by the Sr values. On the other hand Zn loss through heating can not be excluded. Samples with high Zn concentrations therefore strongly support assumptions towards a milk or meat enriched diet.

In general, it is believed that although trace element contents in cremated bones are changed to a certain extent compared to those of not cremated bones, within themselves they seem to be indicative for environmental features.

2. Intraindividual variability

There is a remarkable difference in trace element patterns of samples from one individual (tab. 1, Andernach). The trabecular samples show higher contents than compact parts except the vertebra. The differences might even be as high as several hundred percent (with respect to lead).

The intraindividual variability is found to be higher than variability between individuals (tab. 2, samples Urmitz).

Similar results were reported for not cremated bones some years ago (Brätter et al. 1977; Grupe in press). Therefore also in cremated remains the skeletal part a sample is taken from for trace element analysis should be standardised. The most easy agreement would be to analyse only samples from compact bone, i.e. femoral diaphyses, which are preserved in almost any cremation. Femoral diaphyses have already been used in this paper for interindividual comparison (see below).

3. Interindividual variability

There were seven individuals from Urmitz graveyard tested. The results are remarkably homogenous except for sample 42 : 6. Values for 42 : 6 are either the lowest or highest group limits almost for any element except lead (tab. 1). There are no indications from morphological features or from the Ca/P ratio that mineralization might have been disturbed or cremation was not complete in this specimen. It is assumed that this individual was exceptional for nutritional reasons: the Sr/Zn ratio is very low as is the Sr/Ca ratio. One might suspect a protein rich diet, what is probably supportet by a low Mg level, since Mg is mainly derived from plant products.

Unfortunately, there are no archaeological indicators, i.e. grave goods, that could explain these remarkable differences between 42 : 6 and the other samples in terms of socio-economic features. If one wants to speculate, it should be remembered that Urmitz graveyard belongs to a Roman fortress. Perhaps this individual arrived at Urmitz only a few weeks or months before his death (or simply preferred a protein rich diet). A disadvantage is certainly that 42 : 6 was not deposed in a clay urn but as a cremation layer. So there was no protection from the urn, probably being the reason for a differing trace element pattern. At least Ba can invade the bone and can be bound as barite. But if the Ba value is compared with that from Andernach (deposed in a glass urn!), it seems likely that all values for 42 : 6 are indicative for physiolocial conditions. It would be quite difficult to imagine that during the period of interment Sr and Mg were diminished and Ba, Zn and Pb enriched.

Although we tested only one sample each from two different places than Urmitz (tab. 4), the samples from Andernach and Birkefehl clearly are not located within the Urmitz samples. This is a promising result towards discrimination of groups based on trace element patterns (cf. figs 1 and 2).

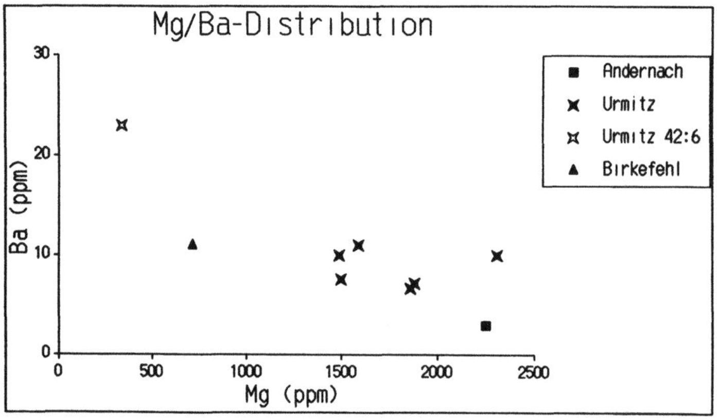


fig. 1: Mg/Ba- distribution of samples. Note location of samples from different sites and exceptional position of Urmitz 42 : 6.

TAB. 4: INTERGROUP-COMPARISON (FEMORA)

| | Ca (mg/g) | P (mg/g) | Sr (ppm) | Sr/Ca | Ba (ppm) | Mg (ppm) | Zn (ppm) | Sr/Zn | Pb (ppm) |
|---|---|---|---|---|---|---|---|---|---|
| Andernach | 252.6 | 138.4 | 116 | 0.46 | 3 | 2250 | 58 | 2.0 | 0.5 |
| Urmitz (excl. 42 : 6) | 349.4 | 162 | 177 | 0.51 | 9 | 1772 | 42 | 4.21 | 1.7 |
| Urmitz 42 : 6 | 255 | 116 | 113 | 0.44 | 23 | 340 | 113 | 1.0 | 2.0 |
| Birkefehl | 389 | 180 | 371 | 0.95 | 11 | 720 | 20 | 18.6 | 0.9 |

Addendum : Quality control by IAEA standard reference material animal bone H5

| element | recommended value | found |
|---|---|---|
| Ca (g/kg) | 211.7 +/- 4.3 | 207.6 |
| P (g/kg) | 95.8 +/- 10.4 | 98.8 |
| Mg (g/kg) | 3.53 +/- 0.05 | 3.58 |
| Sr (mg/kg) | 104.9 +/- 4.9 | 99.2 |
| Zn (mg/kg) | 89.9 +/- 2.3 | 93.0 |
| Ba (mg/kg) | 71.6 +/- 8.9 | 82.5 |

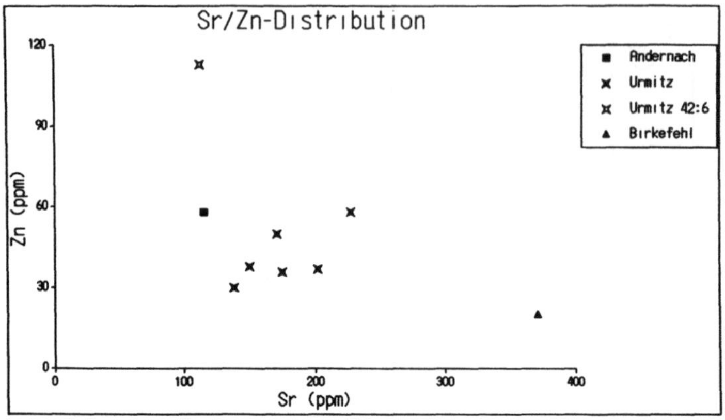


fig. 2: Sr/Zn-distribution of samples. Note location of samples from different sites and exceptional position of Urmitz 42 : 6.

## Conclusions

High temperatures as used for prehistoric cremations obviously do not lead to significant los of those elements which were tested, except Pb. Diagenetic alterations are more or less the same for cremated and not cremated bones, with respect to the mineral compounds.

Further tests should concentrate also on other elements not tested in this paper, as well as on age- or sex related dependencies.

Sr and Zn patterns may be used to reconstruct food chains as in not cremated bones.

There is an increase of Ca due to incineration, which influences the Sr/Ca ratio. Sr/Ca ratios for cremations and not cremated bones are not comparable. A calculation for Sr/Ca ratio in consumed food, which is possible for not cremated bones, is not possible in cremations. A reference series for cremations should be established. It seems reasonable to calculate relative contribution of plants and animals to the food supply from Sr/Zn ratio, at least in addition to Sr/Ca ratio.

Reconstruction of man-environment relations and interactions by trace element analysis of cremations is possible. It is likely that discrimination of population groups can be done by trace element patterning.

References


Brätter P, Gawlick D, Lausch U, Rösick U (1977) On the distribution of trace elements in human skeletons. J Radioanalyt Chem 37: 393-403.

Grupe G (in press) Impact of choice of bone samples on trace element data in excavated human skeletons. J Archaeol Sci.

Hedrich M Communication at the conference.

Hummel S, Schutkowski, H Personal communication.

Iyengar G, Kollmer W, Bowen H (1978) The elemental composition of human tissues and body fluids. Verlag Chemie, Weinheim New York.

Lange M, Schutkowski H, Hummel S, Herrmann B (1987) A bibliography on cremation. Pact 19, Council of Europe, Strasbourg.

Price T, Kavanagh M (1982) Bone composition and the reconstruction of diet:examples from the midwestern United States. Midcontinental J Archaeol 7: 61-79.


**Note added in proof:** In a most recent study on trace elements in ancient human remains (L Runia (1987) The chemical analysis of prehistoric bones. A paleodietary and ecoarcheological study of Bronze Age West-Friesland. BAR Int Ser 363, Oxford) a small subsample of cremations has been analysed too.
This contributes to our still insufficient knowledge on trace element contents in cremations, although first multielement data on cremated human remains were published as early as in 1972 ( B Herrmann (1972) Das Combe Capelle-Skelet. Ausgr Berl 3 : 7 - 69).

## TRACE ELEMENT CONTAMINATIONS IN EXCAVATED BONES BY MICROORGANISMS

Gisela Grupe and Hermann Piepenbrink
Institut für Anthropologie der Georg August-Universität
Bügerstr. 50
3400 Göttingen, FRG

### Introduction

Postdepositional changes of trace element concentrations in excavated human skeletons still remain an unsolved problem in the interpretation of the elemental composition of prehistoric bones. Several physical and chemical factors acting upon the bone mineral during interment are known and have already been subject to experimental approaches (Nelson & Sauer 1984; Lambert et al. 1985; Nelson et al. 1986). But microorganisms invading the corpse after death also play a major part in dead bone decomposition (Baud & Lacotte 1984; Piepenbrink 1986) in being capable for the total destruction of a bone's microstructure by tunneling, even in macroscopically intact specimens (fig. 1). Biogenous decomposition can often be detected by histological methods only and is therefore sometimes overlooked.

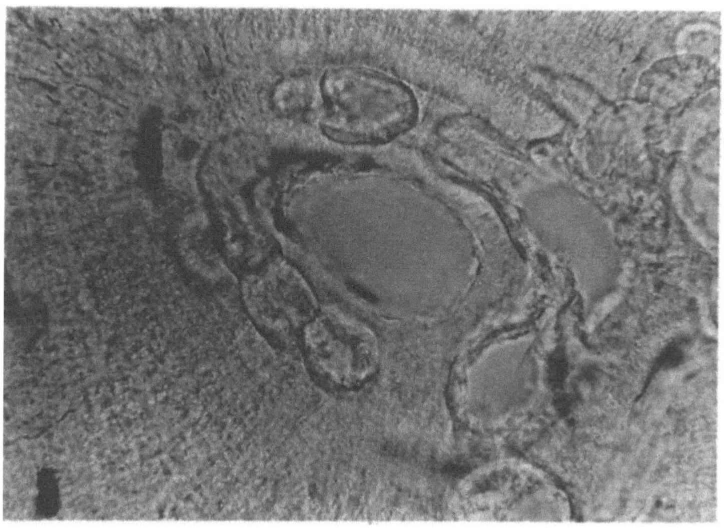

fig. 1: Tunneling caused by microorganisms. Medieval human femurshaft. Magnification: 460x

Microorganisms are destroying the bone collagen, but they may affect the bone's trace mineral content as well in two ways: 1) Microbial activity is the first step in dead bone decomposition, as acid metabolites of the microorganisms are dissolving the hydroxyapatite. Thus, Ca is released from the bone mineral. Soil-pH then plays it's role in either stabilising or further dissolving the mineral (White & Hannus 1983). Thus, microbial activity may not only cause a depletion of those trace elements bound to the bone's collagen, but also of those elements bound to the hydroxyapatite. 2) As microorganisms, especially fungi, are known to accumulate heavy metals, they may also cause a significant contamination by transporting these metals into the bone.

In order to assess the way how microorganisms do affect the trace element content of excavated bones and to get a first impression about the magnitude of such changes, a pilot study was carried out by inoculating fresh bone specimens with microorganisms regularly extractable from prehistoric bones.

## Material and methods

The microorganisms chosen for this study are fungi of the species Penicillium brevicompactum and Cladosporium spec. According to the abbreviations used in our laboratory, these organisms are referred to as "fungus H" and "fungus K" respectively. Both fungi can regularly be detected in excavated bones and are known from previous studies to grow fast, thus invading a bone specimen completely within a few months.

Four femurshafts of freshly slaughtered pigs were cut into three pieces of 2 grams each. The pieces were defleshed in cold water, the fat removed by ether-extraction. They were then sterilised by uranium radiation with a dosis of 2.5 kgy. One piece of each femur was kept under sterile conditions and served as a control. The remaining two pieces were inoculated with the fungus H and K respectively.

In order to test whether the fungi were contaminating the bones with metals from the soil, we simulated "soils" enriched with either Sr, Ba, Zn or Mg, all elements which are supposed to reflect in vivo-conditions in excavated bones, thus being indicators of ancient dietary habits. The inoculated samples were kept on an Oxoid-agar-medium L 11 (Oxoid Deutschland GmbH 1983), which was further enriched with 1% Sr as $Sr(NO_3)_2$, 1% Ba as $Ba(NO_3)_2$, 1% Zn as $ZnSO_4$ and 1% Mg as $MgSO_4$, respectively (fig.2).

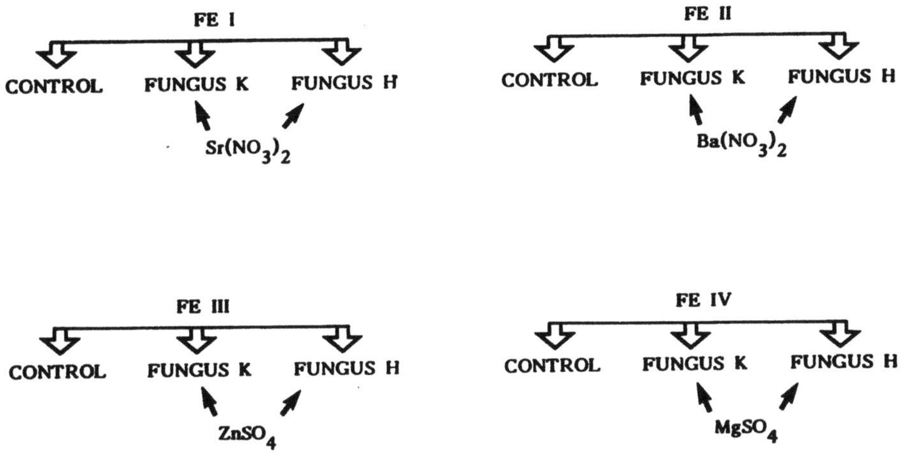


fig. 2: Treatment of the bone specimens.

The inoculated samples were kept for four months in the dark, at aerobic conditions and at room temperature. After the incubation time, the bone specimens were thoroughly cleaned with water using a smooth brush, then kept in an ultrasonic cleaner for 15 minutes. About 2 mm of the surfaces were ground off and the specimens were cleaned again ultrasonically. This cleaning procedure is routinely applied in many laboratories (e.g. Brown & Keyzer 1978). The ultrasonic bath especially is supposed to remove all adherent contaminants from the sample. The surfaces were discarded since the inner parts of compact bone are usually less contaminated than the outer ones (Badone & Farquhar 1982; Lambert et al. 1983; Williams & Marlow 1987).

25 mg of each specimen was wet ashed with $HNO_3$ "suprapur" under pressure for six hours at 160°C and analysed by ICP (Philips spectrometer PO 7450). The IAEA standard "animal bone H-5" (IAEA 1982) was used for calibration. Some material of each specimen was left cleaned, but with intact surfaces to prepare thin sections for investigation by light microscopy.

## Results

The fungi were growing rapidly both on their agar media and on the bone samples, as well superficially as by invasion through natural micropores. Investigation of the thin sections revealed that the bone's microstructure remained unaffected, no tunneling could be observed. The interactions of the microorganisms with the bone were different in so far as the fungus H left no traces in the sample except a dark staining, being intensive at the surface and becoming lighter at the inner parts (fig. 3). This clearly indicates the need of discarding the outer surfaces of the samples. The fungus K was detectable by it's hyphae, penetrating the whole specimen (fig. 4). The hyphae were found within the physiological cavities, clearly indicating that the cleaning method was insufficient since the contaminants were not completely removed.

Trace element analysis revealed differences between the inoculated specimens and the controls (tab. 1). The changes produced by the fungi appeared to be totally unpredictable, since the fungi caused either depletions or enrichments, seemingly independent from the composition of the medium on which they had been kept.

The Ca/P-ratios however, indicating the integrity of the bone matrix, remained unchanged after the incubation (tab. 1), with only one exception: The fungus H, grown on $Ba(NO_3)_2$, had caused a nearly complete depletion of all elements of interest in it's bone specimen. The fungus must have destroyed the bone's mineral matrix, reflected by an extremely deviant Ca/P-ratio. This confirms our previously established hypothesis that gross deviations of the Ca/P-ratio from physiological values in excavated bones indicate excessive alterations of the bone's in vivo-composition due to diagenesis, thus rendering the specimen in question unsuitable for analysis (Grupe 1986).

Specific enrichment of the bone samples by those elements the fungi were especially supplied with are not detectable on the individual level, but are obvious on the group level (tab. 2). Since previous works showed that trace element contents in excavated bones should up to now only be interpreted in a statistical way, we pooled all four controls and used the mean values as the "true in vivo-value" of a small population. Compared with these mean-control-values, the mean trace element content of both inoculated specimens showed specific enrichments with that element the fungi had been especially supplied with. The other specimens incubated on media enriched with other elements than the one in question show trace element contents far closer to the control-values. Remarkably, most of the data are within the range of the control-values.

We then pooled the trace element data of all three samples from one femur, that is we calculated the mean value of the uncontaminated and the contaminated samples

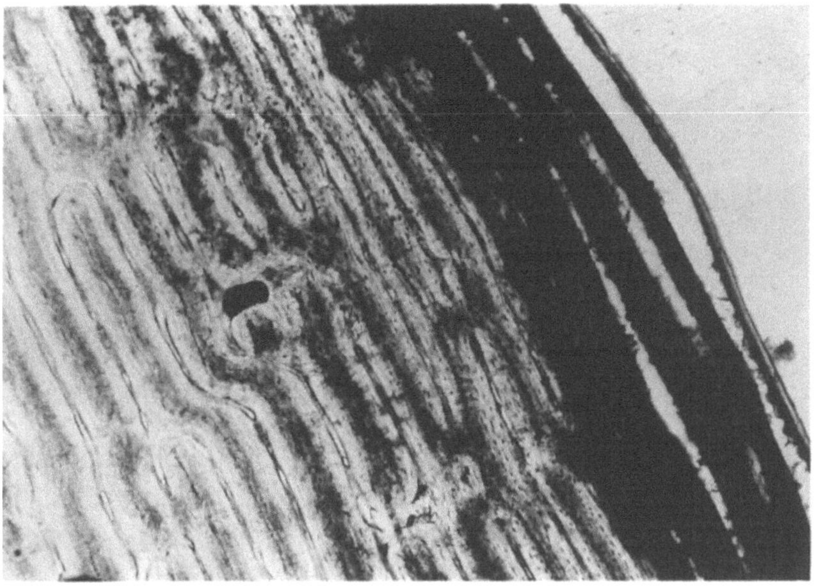

fig. 3: Thin section through the femurshaft of a domestic pig, inoculated with fungus H. Staining is most intensive at the surface, decreasing at the inner parts. Magnification 45x.

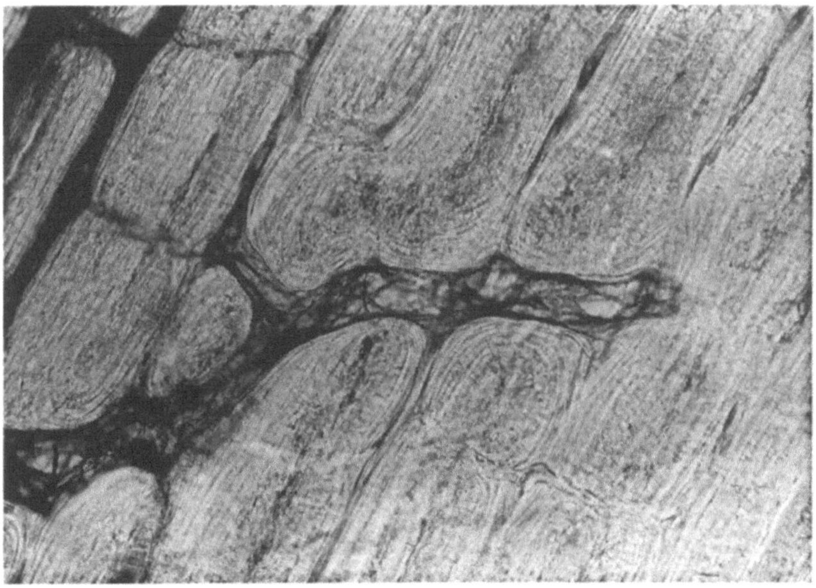

fig. 4: Thin section through the femurshaft of a domestic pig, inoculated with fungus K. The fungus' hyphae are invading the bone through natural micropores. Magnification 110x.

| sample | medium enriched with | fungus | Ca(mg/g) | P(mg/g) | Ca/P | Sr(ppm) | Ba(ppm) | Mg(mg/g) | Zn(ppm) |
|---|---|---|---|---|---|---|---|---|---|
| Fe I | - | - | 296 | 116 | 2.55 | 168 | 8 | 4.88 | 120 |
| Fe I | $Sr(NO_3)_2$ | K | 289 | 118 | 2.42 | 264 | 36 | 4.84 | 164 |
| Fe I | $Sr(NO_3)_2$ | H | 288 | 114 | 2.53 | 440 | 48 | 5.64 | 156 |
| Fe II | - | - | 336 | 122 | 2.75 | 264 | 44 | 6.08 | 124 |
| Fe II | $Ba(NO_3)_2$ | K | 260 | 102 | 2.55 | 248 | 1064 | 4.88 | 156 |
| Fe II | $Ba(NO_3)_2$ | H | 21.6 | 1.2 | 18.0 | 40 | n.d. | 0.52 | 4 |
| Fe III | - | - | 244 | 118 | 2.07 | 76 | 28 | 3.60 | 52 |
| Fe III | $ZnSO_4$ | K | 250 | 116 | 2.16 | 176 | 16 | 4.92 | 156 |
| Fe III | $ZnSO_4$ | H | 330 | 110 | 3.00 | 312 | 60 | 7.08 | 608 |
| Fe IV | - | - | 332 | 132 | 2.52 | 232 | 36 | 5.80 | 120 |
| Fe IV | $MgSO_4$ | K | 392 | 144 | 2.72 | 284 | 52 | 9.60 | 140 |
| Fe IV | $MgSO_4$ | H | 212 | 126 | 1.68 | 52 | n.d. | 2.12 | 112 |

tab. 1: Element contents in compact bone samples from domestic pigs (n.d. = not detected)

| sample | number of specimens | medium enriched with | Ca (mg/g) | P (mg/g) | Sr (ppm) | Ba (ppm) | Mg (mg/g) | Zn (ppm) |
|---|---|---|---|---|---|---|---|---|
| all controls | 4 | - | 302 (244-336) | 122 (116-132) | 185 (76-264) | 29 (8-44) | 5.09 (3.60-6.08) | 104 (52-124) |
| Fe I inoculated | 2 | $Sr(NO_3)_2$ | 287 | 116 | 352 | 42 | 5.24 | 160 |
| Fe II inoculated *** | 2 | $Ba(NO_3)_2$ | 141 | 52 | 144 | 1064* | 2.70 | 80 |
| Fe III inoculated | 2 | $ZnSO_4$ | 290 | 113 | 238 | 38 | 6.00 | 382 |
| Fe IV inoculated | 2 | $MgSO_4$ | 302 | 135 | 168 | 52* | 5.86 | 126 |
| all inoculated | 8 | all | 255 | 104 | 226 | 213** | 4.95 | 187 |

tab. 2: Mean element concentrations in inoculated samples and controls. (* one specimen only, ** six specimens only, *** bone mineral of the specimen inoculated with fungus H deteriorated). Range of the control-values in brackets.

| sample | number of specimens | medium enriched with | Ca (mg/g) | P (mg/g) | Sr (ppm) | Ba (ppm) | Mg (mg/g) | Zn (ppm) |
|---|---|---|---|---|---|---|---|---|
| all controls | 4 | - | 302 (244-336) | 122 (116-132) | 185 (76-264) | 29 (8-44) | 5.09 (3.60-6.08) | 104 (52-124) |
| all Fe I | 3 | $Sr(NO_3)_2$ | 290 | 116 | 291 | 31 | 5.12 | 147 |
| all Fe II *** | 3 | $Ba(NO_3)_2$ | 206 | 75 | 184 | 554* | 3.82 | 95 |
| all Fe III | 3 | $ZnSO_4$ | 275 | 115 | 184 | 35 | 5.20 | 272 |
| all Fe IV | 3 | $MgSO_4$ | 312 | 134 | 189 | 44* | 5.84 | 124 |
| all inoculated | 8 | all | 255 | 104 | 226 | 213** | 4.95 | 187 |

tab. 3: Mean element concentrations, controls and inoculated samples pooled (* two specimens only, ** six specimens only, *** bone mineral of the specimen inoculated with fungus H deteriorated). Range of the control-values in brackets.

and compared the mean with the "true value" calculated before (tab. 3). This procedure needs further explanation since one and the same value (e.g. the Sr-content) now contributes to both mean values with one third and one fourth respectively. But although the sample size is very small, such a comparison may be closer to reality. In a large series of skeletons, the individuals usually show varying degrees of preservation. Whenever trace element analysis is carried out, one would chose the best preserved specimens and not specimens exhibiting obvious signs of deterioration. Therefore, a varying percentage of specimens from a large series is supposed to be well preserved or only slightly decomposed, with trace element contents very close to the in vivo-concentrations. Thus, the mean trace element content of any large series is always calculated from "true" and more or less "changed" data.

The result of this comparison is striking: Almost no differences between the trace element concentrations between the two groups are found except an enrichment of the specific element the fungi had been supplied with (tab. 3).

Pooling all contaminated specimens leads to values much differing from the controls.

## Discussion

Microbial activity in the early stages of bone decomposition obviously does affect the bone's trace element content. These changes are leading either to a depletion or to an enrichment of the specimen with various metals. Fungi are capable to transport soil-elements into the bone. Although our sample size is very small, the results show that induced changes of a bone's in vivo trace element content with respect to Ca, P, Sr, Ba, Mg and Zn are not detectable or insignificant on a group level. Only in such cases where the surrounding medium is markedly enriched with a certain element, this enrichment is also reflected in a significant contamination of the bone. This confirms the need of analysing soil samples adherent to the excavated skeletons, otherwise "unphysiological" trace element contents cannot be judged adequately (Nelson & Sauer 1984). Especially Ba is a contaminating metal in our experiment. Once carried into the bone by a fungus, this element may become bound as barite in the later stages of decomposition. Ba therefore seems to be unsuitable for dietary reconstruction by trace element analysis, perhaps with the exception of a few cases with extraordinarily well preserved skeletons.

It has been stressed previously (Price et al. 1985) that trace element contents of prehistoric bones interred into different soils (e.g. of different geographical origin) cannot be directly compared. This is also confirmed by our findings, as the pooled data of all inoculated specimens on different media show marked discrepancies with the control-values.

Microorganisms are capable to deteriorate both the bone's mineral and it's collagen. Solubilization of the hydroxyapatite by acid metabolites is supposed to be reflected in altered Ca/P-ratios. In all cases but one, this ratio in the inoculated specimens did not differ from the controls, indicating the integrity of the bone's mineral matrix. Since the histological investigation showed intact microstructures but hyphae in the microcavities of the section, we conclude that the contaminating elements in our experiment are still bound to the tissues of the fungi.

Further studies therefore will have to deal with the elaboration of more sufficient cleaning procedures. The chitinous matrix of the hyphae has to be taken into account.

## Conclusion

Trace element analysis of excavated human bones has to be paralleled with soil analysis since fungi invading the bone in the early stages of decomposition are actively carrying chemical constituents of their surrounding medium with them, thus causing a contamination of the bone specimen. This active transport is independent from any chemical of physical gradient. Solubilization of the bone mineral by the microorganism's metabolites may also cause leaching of some elements out of the bone.

But when soil composition is taken into account, trace element loss or contamination induced by microorganisms seem to be insignificant on a group level. These findings are very similar e.g. to palaeoserological results.

Cleaning procedures have to be worked out to remove organic residues from the bone's micropores.

Tunneling detectable in thin sections of excavated bones may be a phenomenon of a later stage of decomposition and is therefore not visible in our specimens. Until sufficient cleaning methods are not available, a histological investigation must precede every elemental analysis. We recommend to discard all specimens showing extensive tunneling or loss of collagen, detectable by a loss of birefrigence, as both phenomena are indicators for a previous invasion by microorganisms.

## Acknowledgement

We are most grateful to Dr. R. Ender and his staff from the Philips ICP-Laboratory in Fellbach for providing us with the measurements.


## References


Badone E, Farquhar RM (1982) Application of neutron activation analysis to the study of element concentration and exchange in fossil bones. J Radioanalyt Chem 69: 291-311.

Baud C-A, Lacotte D (1984) Etude au microscope électronique à transmission de la colonisation bactérienne de l'os mort. CR Acad Sc Paris 298: 507-510.

Brown AB, Keyzer H (1978) Sample preparation for strontium analysis of ancient skeletal remains. Contribs Geol 16: 85-87.

Grupe G (1986) Multielementanalyse: Ein neuer Weg für die Paläodemographie. Materialien Bevölkerungswiss, Sonderheft 7, Wiesbaden.

IAEA (1982) International Atomic Energy Agency Progress Report No 1: Intercomparison of minor and trace elements in IAEA animal bone (H-5), Wien.

Lambert JB, Simpson SV, Buikstra JE, Hanson D (1983) Electron microprobe analysis of elemental distribution in excavated human femurs. Amer J Phys Anthrop 62: 409-423.

Lambert JB, Simpson SV, Weiner SG, Buikstra JE (1985) Induced metal-ion exchange in excavated human bone. J Archaeol Sci 12: 85-92.

Nelson DA, Sauer NJ (1984) An evaluation of postdepositional changes in the trace element content of human bone. Amer Antiquity 49: 141-147.

Nelson BK, DeNiro MJ, Schoeninger MJ, De Paolo DJ, Hare PE (1986) Effects of diagenesis on strontium, carbon, nitrogen and oxygen concentration and isotopic composition of bone. Geochim Cosmochim Acta 50: 1941-1949.

Oxoid Deutschland GmbH (1983) Handbuch der "Oxoid"-Erzeugnisse für mikrobiologische Zwecke, Wesel.

Piepenbrink H (1986) Two examples of biogenous dead bone decomposition and their consequences for taphonomic interpretation. J Archaeol Sci 13: 417-430.

Price TD, Schoeninger MJ, Armelagos GJ (1985) Bone chemistry and past behavior: an overview. J Hum Evol 14: 419-447.

White EM, Hannus LA (1983) Chemical weathering of bone in archaeological soils. Amer Antiquity 48: 316-322.

Williams CT, Marlow CA (1987) Uranium and thorium distributions in fossil bones from Olduvai Gorge, Tanzania and Kanam, Kenya. J Archaeol Sci 14: 297-309.

# TRACE ELEMENT ANALYSIS OF HUMAN HAIR

K.Dörner
University Children's Hospital
Schwanenweg 20
2300 Kiel 1
FRG

Trace element analysis of hair has met a considerable interest in nutritional sciences, in clinical and in forensic medicine. 25 years ago, Swedish and British scientists attracted attention by the finding of high arsenic concentrations in consecutive hair sections of emperor Napoleon's hair (Smith et al. 1962). They suspected that Napoleon died from a chronic arsenic intoxication induced over a period of at least 4 months. Another startling example of hair analyses in environmental history is derived from a bird, from the goshawk which lives on small seed eating birds (Thomson 1977). Johnels and Westermark determined the mercury content of feathers over a period of one hundred years back to 1860's mainly based on museum exhibits of female goshawks. They found a constant level from the year 1863 to 1946 of about 2.2 µg/g. Up to 1966 a dramatical increase was found with a mean value of 29 µg/g. This increase was explained by the use of methylmercury compounds as fungicides for seeds in the period from 1940 to 1966. When the application of methylmercury was prohibited in 1966 the mercury contents of the feathers quickly decreased to the previous level.

These two examples illustrate, why hair analyses are popular in medicine, nutrition and environmental science. Table 1 summarizes the advantages of hair sample testing. The last item of Table 1, however, has been criticized as it will be shown later. Trace element analysis of hair has widely been used in animals and men. In human hair it was applied to approach many fields, which are listed in Table 2.

In diabetic children (partly before treatment) lower zinc (Amador 1975a; Pleban 1980; Dörner 1985), chromium (Hambidge 1968; Pleban 1980) and manganese (Pleban 1980) levels were described in comparison to healthy controls. These findings should certainly not be regarded as a new alternative pathway in diabetes diagnostics, but they could reveal an insufficient supply with these metals by increased renal losses, as it has been known for a long time. These results are the more remarkable as even etiological connections have been drawn between diabetes and a deficiency of these trace metals.

Tabel 1. Advantages of testing hair samples

1. Easy material collection (atraumatic)
2. Structural and chemical stability (transport, storage)
3. Trace element concentrations are high in relation to blood or urine
4. Chemical composition averages long term body supply
5. Intraindividual longitudinal studies are possible
6. Trace element concentrations correlate with endogenous supply and/or in situ environmental pollution

Acrodermatitis enteropathica is an inborn error of metabolism with strongly decreased intestinal zinc absorption and extremely low serum zinc levels. Low hair values as found by Amador (1975b), Nelder (1975) and later by Walravens (1978) are therefore not surprising. Hair zinc concentration raises after oral supplementation with zinc (Nelder 1975).

Table 2. Diagnostic use of hair analyses

Laboratory Diagnostics of Diseases
- Diabetes (Zn, Cr, Mn)
- Acrodermatitis enteropathica (Zn)
- Cystic Fibrosis (Zn)
- Morbus Wilson and Menkes Disease (Cu)
- Keshan Disease (Se)

Nutritional Status Studies
- Growth Retardation (Zn)
- Pregnancy and Fetal Malformations (Zn, Cu, Cr, Mn)
- Population living on Se-exhausted soils

Toxicological Studies
- Environmental pollution (Hg, Pb, Cd)
- Haemodialysis (Al)

We have found inconsistent results using hair analysis in another disease, where malabsorption of zinc is a general finding later leading to low serum zinc values: In cystic fibrosis (CF) we found significantly lower copper and zinc hair values in a large collective of CF patients compared to healthy school children (Dörner 1985), which confirms the findings of Jacob (1978), whereas others (Mitchell 1980,

Gyurkovits 1978) found no differences between growth retarded CF patients and controls.

Two diseases of copper metabolism, M.Wilson and Menkes`Disease present low serum copper concentration for different reasons and, thus, consequently suggest low hair concentrations. Data from the literature are ambiguous. We found hair copper below the $P_{15}$ in 1 of 4 patients with M. Wilson and in 3 of 4 patients withs Menkes` Disease (Dörner 1985), whereas Rice and Goldstein (1961) got low hair concentrations only in females, but not in male patients. Bradfield (1980) denies any usefulness of hair analysis in copper deficiency.

Keshan Disease is a cardiomyopathy highly correlated with low selenium intake. Up to this day, this disease was only found in a region of China with an endemic low selenium concentration of soil. Hair selenium contents are below 0.12 ppm (nonendemic areas: > 0.16 ppm). Keshan Disease is thus associated with low hair selenium (Yang 1985).

Nutritionists performing field studies are in urgent need of easily accessible test materials. Hair would be ideal for them. Growth retardation is a common sign of many types of undernutrition, which need a corroboration by chemical analysis. Unfortunately, earlier hopes for hair zinc as a biochemical indicator of growth retardation and zinc deficiency (Hambidge 1972) have been disappointed recently (Dorea 1985). One of the problems seems to be the dependency of hair growth from nutrition and health itself, as Hambidge (1980) and Bradfield and Hambidge (1980) pointed out. In severe malnutrition children`s hair growth slows down considerably and hair zinc may become "normal" or even high (Table 3). Thus hair zinc has a limited value as a nutritional index (Shapcott 1982). In recent years, it has been continuously discussed whether blood zinc and hair zinc are correlated (ideally : linearily) and, furthermore, if no correlation is found, whether this fact would limit the usefulness of hair analysis. Indeed, we found no correlation between hair zinc and hair copper and the corresponding serum values in a study with a homogenous group of 591 healthy school children (Dörner 1985). Similar results were presented by others (Mc Bean et al 1971, Vir et al 1979). Advocates of hair analysis argue that these parameters cannot correlate, because blood concentrations fluctuate quickly in comparison to hair concentrations (Laker 1982, Medeiros 1982). These authors consider hair analysis useful for longterm studies.

Table 3. Laboratory Assessment of Zinc Nutritional States (Hambidge 1980)

| Zinc Nutritional State | Hair Growth | Plasma Zinc | Hair Zinc |
|---|---|---|---|
| normal | normal | normal | normal |
| mild deficiency | normal | normal | decreased |
| severe deficiency | decreased | decreased | normal |

Hair concentration of zinc, copper, chromium and manganese were determined in the hair of pregnant women with the intention to find correlations with number of pregnancies, serum values, age or even fetal malformations. The results obtained were not statistically significant and if there were any, they do not convince (Vir et al 1981, Shapcott et al 1980, Sanner et al 1985). Summarizing the nutritional aspects of hair analysis it may be concluded, that hair analysis is a more or less limited and improper tool (Rivlin 1983) with the only exception of the hair selenium contents in low selenium areas in China mentioned earlier.

In toxicological analyses concerning elements which normally show very low concentrations in human tissues as lead, mercury, and cadmium, hair analysis may be even more indicative than blood analysis (Laker 1982), and in ancient samples hair is - besides bone - the only available human material. Examples from this field were given at the beginning. Mercury and cadmium levels are higher in hair than in blood (Clarkson 1976, Berg and Kollmer 1983). Similarly, arsenic poisoning may be suspected by hair analysis (Underwood 1977). Finally, the high aluminium concentrations regularly found in the blood of patients with haemodialysis are expressed in the hair of these persons (Marumo et al 1984). In comparison with nutritional status studies and the chemical pathology of the disease mentioned, it may be concluded that hair analysis is most valuable in toxicological investigations.

A final judgement of hair analysis cannot be done without enumerating the pitfalls of this type of analysis (Taylor 1986), as summarized in Table 4.

Table 4. Limitations and pitfalls in hair analysis

1. Cosmetic treatment and environmental factors
2. Sampling techniques (location, length)
3. Preanalytical washing procedure
4. Sample ashing and determination
5. Reference values and interpretation

The chemical composition of the bradytrophic tissue hair is subjected to numerous exogenic influences in vivo: dyeing, bleeching, shampooing, even showering with water with a high copper content have been shown to increase trace element concentration (Clanet et al 1982, Schroeder and Nason 1969). On the other hand, a decrease of trace elements in the outer layers of hair by detergents should be possible, too. Creason et al (1975) studied the influence of environmental exposure on scalp hair trace element levels in 3 areas of New York with different pollution exposure. In many elements significant positive correlations between trace elements in hair and dustfall were found.

Hair analysis should be performed with samples taken closely from the scalp to diminish exogenous influences. It can not be recommended to study single hairs, as some hairs may be in the resting and more or less in the growing phase, and interpretation of the results might get difficult. Sections of 20 or more different hairs give a more reliable average.

Trace elements in pubic hair were compared with those in scalp hair under the impression that cosmetic and/or environmental factors might be less interfering in pubic hair with the endogenous content (De Antonio 1982, Shapcott et al 1980). To date, however, the use of pubic hair does not seem to provide any advantage.

Hair samples have to be cleaned from adhering epithelial cells, sweat and external dirt. Numerous washing procedures have been proposed and they obviously remove different amounts of trace elements from the hair samples. In general, water and organic solvents like aceton, ethanol and diethylether are used (no ionic detergents!).

The sample ashing technique should guarantee that no trace elements are volatilized (Friel and Ngyen 1986). Dry ashing is not recommended, as many elements are volatile and might be lost. Accuracy and precision of the analytical technique must suffice the state of the art.

The major problem of hair trace element analysis are reference values and the interpretation of the results. Many authors have described age, sex and hair colour related dependencies, just as many have denied them. Apart from methodical and statistical problems which complicate the comparison of results from different laboratories, it would be best if each laboratory which performs hair analysis provides reference values of its own. Even in this case the range of hair values of healthy persons is large in comparison with serum values and the interpretation of borderline values is difficult (Table 5).

The only purpose for which hair analysis are still as an additional - not as an isolated! - parameter in clinical diagnostics is the fact, that reliable criteria are lacking for the definition of a trace element deficiency.

Table 5. Median, 80% range and median-related range (% range) of serum and hair concentrations of zinc and copper in healthy school children (adtd from Dörner 1985)

| | Zinc | Copper |
|---|---|---|
| serum (mg/l) | 0.825 (0.713 - 0.988) | 0.975 (0.765 - 1.19) |
| hair (mg/kg) | 200 (101 - 452 ) | 31.0 (12.0 - 110 ) |
| % range serum | 33.3 | 43.6 |
| % range hair | 176 | 316 |

Commercially available hair analyses by profit organisations have got well appreciated by alternative medicine doctors as well in Europe as in the U.S.A. for diagnosis of deficiencies and diseases. Without information on all the factors listed above, no serious interpretation is possible, and suggestions for a supplementation therapy are quackery (Sherertz 1985).

Hair from archaeological specimens was supposed to be a valuable source of information on nutrition and health of ancient populations by Benfer and coworkers (1978), who analyzed 8 hair samples from Peruvians most of which were several thousands years old. One of their final conclusions was that these persons had not suffered from diabetes, because their hair chromium was high. As no accompanying soil samples were analyzed, this seems to be rather hypothetical regarding all the pitfalls of hair analysis cited above. At the moment analysis of ancient hair is an interesting, but solely experimental matter. Our preliminary study with hair samples from the 11th, 12th and 18th century should be understood in this way. The samples were given to us by Dr.Gisela Grupe, Göttingen. The older ones are from Schleswig (abbreviation: RM), the younger ones from Braunschweig (abbreviation: ST). The outer appearence of the cleaned samples is shown in Fig.1,2. For trace element analysis the hair samples were separated as well as possible from adhering soil and washed by our standard procedure published recently (Dörner 1985). After wet pressure ashing the trace elements were determined by atomic absorption spectrophotometry (AAS). The results are given in Figs. 3 - 7 where hair (line), (soil) residue (R) and scalp (S) values are given together with nowaday reference values. The main findings are:

1. As a rule, hair concentrations correlated with the concentrations in the residues or the scalp which are higher nearly in all cases.
2. The metal concentrations of the ancient hair is comparable to those from our days.
3. The group samples from Braunschweig (all females, ST) has low calcium, strontium

and zinc values; copper is very high in 3 of 4 cases. The soil concentration parallels the hair concentration, however.

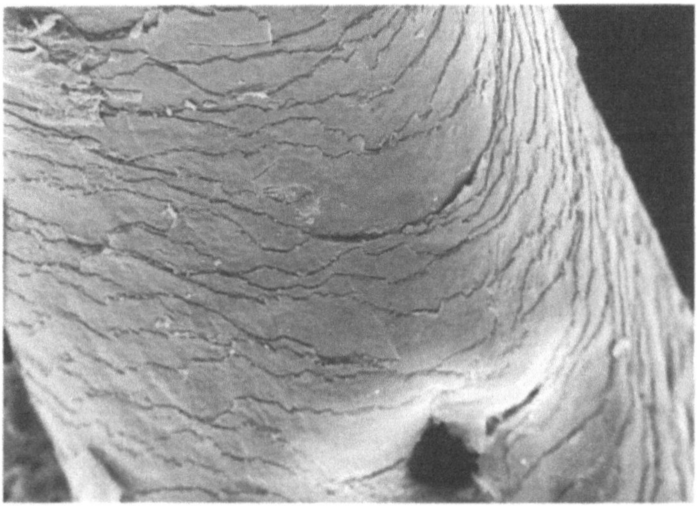

Fig. 1. Electron microscopy (SEM) of an 18th century hair sample ( female, ST 13) Magnification 1150

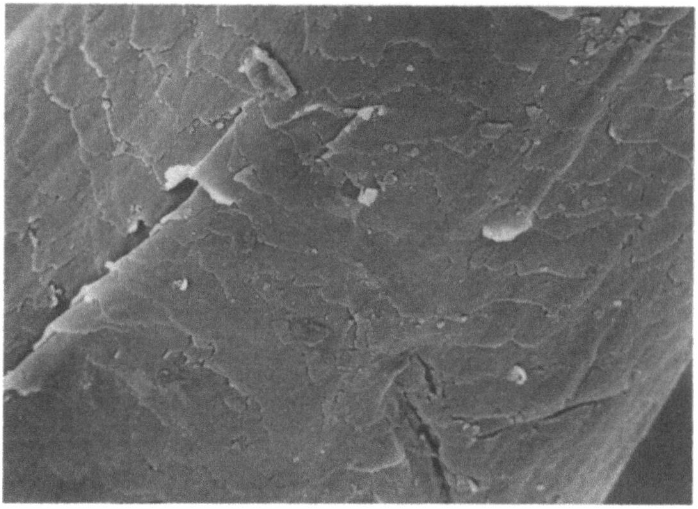

Fig. 2. Electron microscopy (SEM) of an 11th/12th century hair sample (child,RM 22) Magnification 1100

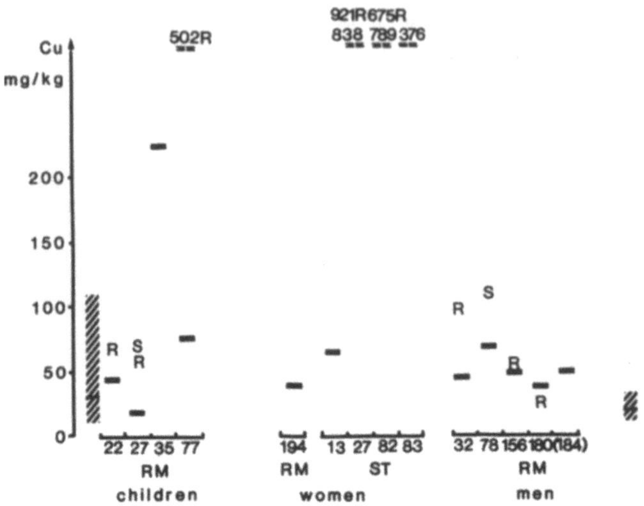


Fig.3. Copper concentrations in the hair (-), in scalp (S) and in hair adherent residue (mostly soil, R) of samples from Schleswig (abbreviation: RM with code number) and Braunschweig (abbreviation: ST with code number). Results outside the scale of the figure are given by their exact values at the upper margin. The hatched bar graphs present nowaday ranges of comparison values for children (left, Dörner 1985) and adults (right, Iyengar et al 1978). Sample RM 184 is from an adult individual of undetermined sex.

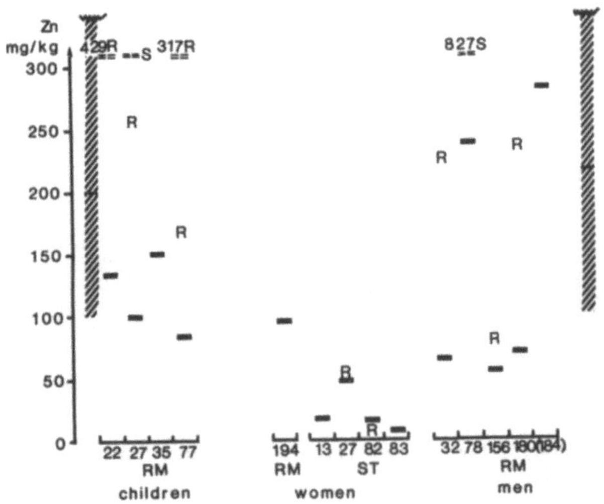


Fig.4. Zinc concentrations in the hair (-), in scalp (S) and in hair adherent residue (mostly soil, R) of samples from Schleswig (abbreviation: RM with code number) and Braunschweig (abbreviation: ST with code number). For further explanations see Fig. 3.

There is an urgent need to elucidate the metal transfer from soil to hair and also vice versa before drawing any conclusions on the nutritional status or on health of the ancient probands. This could be managed by laboratory trials with close contact of hair with different soil samples at high humidity and different temperatures , as well as by extrinsic radioactive marking of soil samples and consecutive examination of the metal uptake by the hair. The reverse exchange processes could perhaps be studied by neutron activation of trace metals in hair, followed by long term exposition of these activated hair samples to soil and follow-up of long-life isotopes like $^{65}Zn$ or $^{60}Co$. An alternative pathway would be feeding rats with $^{65}Zn$ and studying the carry-over of hair zinc to soil samples.

Acknowledgement: The electron microscopic photographs were kindly provided by Ms Müller from the Anatomy Department, University of Kiel (Dir.: Prof.Dr.Tillmann)

## References

Amador M, Pena M, Garcia-Miranda A, Gonzalez A, Hermelo M (1975a) Low hair-zinc concentration in acrodermatitis enteropathica. Lancet 1379

Amador M, Hermelo M, Flores P, Gonzales A (1975b) Hair zinc concentrations in diabetic children. Lancet 1146

Benfer RA, Typpo JT, Graf VB, Pickett EE (1978) Mineral analysis of ancient peruvian hair. Am J Phys Anthrop 48: 277-282

Berg D, Kollmer WE (1983) The deposition of inorganic mercury in newly grown hair of the rat. In: Brätter P, Schramel P (eds) Trace Element-Analytical Chemistry in Medicine and Biology, Vol 2, Walter de Gruyter & Co. Berlin, 571-582

Bradfield RB, Cordano A, Baertl J, Graham GG (1980) Hair copper in copper deficiency. Lancet 343-344

Bradfield RB, Hambidge KM (1980) Problems with hair zinc as an indicator of body zinc status. Lancet 363

Clanet P, DeAntonio SM, Katz SA, Scheiner DM (1982) Effects of some cosmetics on copper and zinc concentrations in human scalp hair. Clin Chem 28: 2450-2451

Clarkson ThW (1976) Quantitative measures of the toxicity of mercury in man. In: Prasad AS, Oberleas D (eds) Trace elements in human health and disease, Vol II Essential and toxic elements. Academic Press, New York, 453-475

Creason JP, Hinners ThA, Bumgarner JE, Pinkerton C (1975) Trace elements in hair, as related to exposure in metropolitan New York 21: 603-612

DeAntonio SM, Katz SA, Scheiner DM, Wood JD (1982) Anatomically-related variations in trace-metal concentrations in hair. Clin Chem 28: 2411-2413

Dörner K (1985) Die Bestimmung von Spurenelementen und ihre klinische Bedeutung Enke, Stuttgart

Dorea JG, Paine PA (1985) Hair zinc in children: Its uses, limitations and relations and relationship to plasma zinc and anthropometry. Hum Nutr:Clin Nutr 39C: 389-398

Friel JK, Nguyen ChD (1986) Dry- and wet-ashing techniques compared in analyses for zinc, copper, manganese, and iron in hair. Clin Chem 32: 739-742

Gyurkovits K, Rutishauser M (1978) Cinkanyagcsere valtozasok mucoviscidosisban (mv) (Hung.: Changes of zinc metabolism in mucoviscidosis) Gyermekgyogyaszat 29: 157-160

Hambidge KM (1980) Hair analyses. Pediatr Clin North Am 27: 855-860

Hambidge KM, Hambidge C, Jacobs M, Baum JD (1972) Zinc in hair. Pediat Res 6: 868-874

Iyengar GV, Kollmer WE,Bowen HJM (1978) Trace elemental composition of human tissues and body fluids. Verlag Chemie, Weinheim

Jacob RA, Sandstead HH,Solomons NW, Rieger Ch, Rothberg R (1978) Zinc status and vitamin A transport in cystic fibrosis. Am J Clin Nutr 31: 638-644

Laker M (1982) On determining trace element levels in man: The use of blood and hair. Lancet II: 260-262

Marumo F, Tsukamoto Y, Iwanami Sh, Kishimoto T, Yamagami S (1984) Trace element concentrations in hair, fingernails and plasma of patients with chronic renal failure on hemodialysis and hemofiltration. Nephron 38: 267-272

McBean LD, Mahloudji M, Reinhold JG, Halsted JA (1971) Correlation of zinc concentrations in human plasma and hair. Am J Clin Nutr 24: 506-509

Medeiros D (1982) Trace elements in human hair. Lancet II: 608

Mitchell EA,Huymans M, Elliott RB (1980) Serum and hair zinc in cystic fibrosis NZ Med J 92: 6-7

Neldner KH, Hambidge KM (1975) Zinc therapy of acrodermatitis enteropathica N Engl J Med 879-882

Pleban PA, Matsushige L (1980) Determination of chromium, copper, manganese and zinc in hair from apparently healthy individuals and diabetics. Clin Chem 26: 1055

Rice EW, Goldstein NP (1961) Copper content of hair and nails in Wilson's disease (hepatolenticular degeneration). Metabolism 10: 1085-1087

Rivlin RS (1983) Misuse of hair analysis for nutritional assessment. Am J Med 75: 489-493

Saner G, Dagoglu T, Özden T (1985) Hair manganese concentrations in newborns and their mothers. Am J Clin Nutr 41: 1042-1044

Schroeder HA, Nason AP (1969) Trace metals in human hair. J Invest Dermatol 53: 71-78

Shapcott D (1982) Hair and plasma in the diagnosis of zinc deficiency. In: Clinical applications of recent advances in zinc metabolism, Alan R Liss New York, 121-139

Shapcott D, Cloutier D, Demers OO, Vobecky JS, Vobecky J (1980) Hair chromium at delivery in relation to age and number of pregnancies. Clin Biochem 13: 129-131

Sherertz EC (1985) Misuse of hair as a diagnostic tool.Arch Dermatol 121: 1504-1505

Smith H, Forshufvud St, Wassen A (1962) Distribution of arsenic in Napoleons hair. Nature 194: 752-726

Taylor A (1986) Usefulness of measurements of trace elements in hair. Ann Clin Biochem 23: 364-378

Thomson AJ (1977) Giftige Elemente in der Umwelt. Nachr Chem Techn Lab 25: 708- 710

Underwood EJ (1977) Trace elements in human and animal nutrition. 4th ed. Academic Press, New York

Vir ShC, Love AHG (1979) Zinc and copper status of the elderly. Am J Clin Nutr 32: 1472-1476

Vir ShC, Love AHG, Tompson W (1981) Serum and hair concentrations of copper during pregnancy. Am J Clin Nutr 34: 2382-2388

Walravens PhA, Hambidge M, Neldner KH, Silverman A, van Doorninck WJ, Mierau G, Favara B (1978) Zinc metabolism in acrodermatitis enteropathica. J Pediatr 93: 71-73

Yang GQ (1985) Keshan Disease: An endemic selenium-related deficiency disease. In: Trace elements in nutrition of children. Raven Press, New York, 273-290

# THE HEAVY METAL BURDEN IN ANCIENT SOCIETIES

Tony Waldron
31 Maidstone Road
Bounds Green
London N11 2TR, UK

There is no entirely satisfactory definition of what constitutes a 'heavy metal' but the group is generally taken to include at least the following: silver (atomic weight 108), cadmium (112), tin (119), antimony (112), platinum (195), gold (197), mercury (201), thallium (204) and lead (207). In their various ways, these metals have all been extremely important in human history because of their technological or commercial value. Four of the group (cadmium, lead, mercury and thallium) are toxic to man and all but thallium might have been encountered in different ways by peoples in antiquity.

Exposure to toxic metals may be environmental or occupational (or both) and this distinction is important when assessing the extent to which they might have constituted a health hazard in the past. For mercury, exposure would have been predominantly occupational, for cadmium it would have been environmental whilst for lead, it would have been both.

<u>Mercury:</u> Mercury was well known in antiquity and cinnabar from Almaden in Spain was traded by the Phoenaecians. It was used as the pigment vermillion for decorative purposes and also as a cosmetic. The mercury mines at Almaden were notoriously unhealthy and they were worked in Roman times (as were the siiver mines) by prisoners or slaves, almost all of whom would have developed mercury poisoning from inhaling mercury vapour present in the atmosphere of the mine. In the middle ages the illuminators of manuscripts used mercury to apply gold to their work. An amalgam of gold and mercury was made and painted on the parchment or vellum and the mercury allowed to evaporate, with or without application of heat, leaving the gold behind. This very dangerous process was used on a much larger scale in later times for applying gold decoration to furniture or to small objects such as buttons; a similar process was used to silver mirrors. Almost all those who were engaged in this type of work could expect to be poisoned by it either by inhaling mercury vapour or by allowing the mercury to come into contact with their skin. There was little release of mercury into the general environment, however, and those who did not work with it would have been

unlikely to have come into contact with it. This is the reverse of the modern situation in which occupational mercury poisoning is now rare (at least in the developed countries) but endemics of environmental poisoning are not infrequent because of the use of mercurial fungicides to treat grain and the release of mercury in factory effluent into rivers and lakes from which fish are taken as the main source of food by the indigenous population.[1]

Cadmium: Cadmium was almost never used in ancient technology except in the manufacture of an unusually glossy bronze of bright yellow colour and it is essentially a 'modern' metal. It is possible that some environmental exposure might have occurred, however, to those living on soils naturally rich in cadmium or who used cadmium-rich water for irrigation or drinking. Since some lead ores also contain cadmium, exposure to cadmium might have occurred where such ores were in use. Modern experience would suggest, however, that none of these exposures would have been likely to be associated with serious effects on health since high body burdens, such as occur with occupational exposure, are required before harmful effects on the kidneys and other organs become apparent.[2]

Lead: Lead has been used by man in one form or another for at least 4.000 years although originally it was valued not for its own sake, but because it was the principal source of silver in the ancient world and the silver mines of antiquity were, in fact, galena (lead sulphide) mines. The most important of these mines were in Asia Minor and the ores from some of these contained in excess of 1.5% of silver. It was on mines such as this, and especially the one at Laurion, that the wealth of Athens largely depended. Silver was recovered from the ore by the process of cupellation which was an exceedingly dangerous task. In essence the process was simple; the ore was melted and the lead oxide which formed on the surface of the molten metal was removed until finally, silver of varying degrees of purity remained. Lead oxide is a highly toxic form of the metal because of its great solubility in tissue fluids and it is impossible to believe that anyone involved in this work would have escaped unharmed. It was actually much more dangerous than mining the galena which, being almost insoluble, is relatively non-toxic.

Lead itself was extensively used by the Roman water engineers and although this practice was condemned by Vitruvius[3] it would have been hazardous only in those areas in which the water was soft or acidic because only under these conditions is the water plumbo-solvent. Much more dangerous was the accidental or deliberate adulteration of food and drink. And of these, the most hazardous was the addition of lead to wine to improve its taste and hence its saleability. This practice continued into quite modern times and ensured that a substantial proportion of the population was exposed to lead in some degree or other and that lead poisoning was also common[4].

## The determination of heavy metal exposure

The three heavy metals with which we are concerned here tend to accumulate in the body and the total body burden of each can, in theory, be used to assess exposure on either an individual or a population basis. There are difficulties in undertaking this assessment in past populations, however, because of differences in their distribution within the body. Cadmium and mercury accumulate preferentially in the kidney and to a lesser extent in the liver and it is these organs which contain the greater part of the body burden. Their concentration in skeletal tissues is very low, usually less than one part per million (ppm), and this is probably not an accurate reflection of the total body burden. On the other hand, because it is metabolised in the body in a manner similar to calcium, the concentration of lead is higher in the skeletal than in the soft tissues and the bones and teeth contain at least 90% of the total body burden. This makes the determination of exposure to lead in the past much easier than either that of cadmium or mercury since there is abundant skeletal material on which to work. It is with lead exposure, therefore, that the remainder of this chapter will be largely concerned.

## Determination of lead exposure in past populations

There are some important features of lead metabolism which need to be considered when attempting to determine exposure in past populations. Firstly, the concentration of lead in the skeleton is not constant but varies not only from bone to bone, but also from site to site in a single bone.[5] Therefore the sampling procedure must be standardised so that the portion for analysis is taken from the same point on the same bone. Only if this is done can valid intra- or inter-study comparisons be made. Similarly, the concentration in the teeth varies significantly so that if the determination of exposure is to be based on the amount of lead in the teeth, the same tooth must be used in all cases.[6] Secondly, the concentration of lead in the skeleton increases with age until about the 5th or 6th decades when it may decrease as the skeleton becomes relatively demineralised. Thirdly, modern studies have shown that, between the ages of about 15-50, the blood lead concentration in women is lower than in men and hence the bone lead concentrations will be similarly affected.[7] Studies on past populations should reflect these important age and sex differences; if they do not, then some explanation must be called for. Finally, bone lead concentrations in a population follow a normal distribution which is skewed to the right; data from ancient material must also approximate to such a distribution before they can be considered epidemiologically reliable.

Whenever the conditions referred to above are not met, it is highly unlikely that the lead levels found truly reflect the exposure of the population from which the samples were taken. And the most likely explanation is that the bones have taken up lead post-mortem.

## Post-mortem uptake of lead

There is not the slightest doubt that, under certain conditions, bones can take up lead from the surroundings in which they are buried.[8] The most obvious case is when the body has been buried in a lead coffin (vide infra) but contamination from the soil can also be so great as to render chemical analysis almost worthless. This applies equally to trace elements other than lead and as a general rule, wherever any elemental analyses are planned to be carried out on bone, soil samples would also be taken for analysis. Where post-mortem contamination is suspected or proved, this can be compensated for either by determining the lead content of deep layers of bone by removing the outer few 100 µm of the sample prior to analysis, by 'coring' out a sample from within the cortex of, for example, the femoral shaft, or by using physical methods of analysis to determine the concentration at different depths from the surface. If these methods cannot be used, an estimate of the mean lead content of the sample which is not due to soil contamination can be derived by regression analysis. For example, bones from a site at Borderley Abbey in Worcestershire were found to be highly contaminated with lead from the soil so that their mean lead content was 682 µg/g dry weight. This is a level which is biologically entirely implausible. From the regression equation of bone lead on soil lead, it was found that 59% of the variance was due to contamination and that the mean 'uncontaminated' concentration was ca 40 µg/g, a level which was entirely convincing.[9]

## Lead exposure at the Romano-British site at Poundbury

Results from bones obtained from the Romano-British cemetery at Poundbury in Dorset may be used to illustrate the extent of lead exposure at that time. Soil samples were taken from a sub-set of the burials and showed that there was little correlation between bone lead and soil lead concentrations and that the variance in bone lead concentrations which was due to lead in the soil was less than 9% ($R^2 = 0.089$).[10] A total of 336 analyses was undertaken (using rib samples in all cases) and the range of lead concentrations was from 12-449 µg/g dry weight. The concentrations were normally distributed with a tail to the right and there was a small but statistically significant difference between the concentrations in males and females (table 1); there was also an increase in lead level with age, from 15 up to the age of about 55 (table 2).

| | Mean | SD | Median | Range | n |
|---|---|---|---|---|---|
| Males | 118.9 | 56.8 | 110.5 | 40-319 | 130 |
| Females | 102.8 | 60.2 | 90.5 | 12-449 | 138 |
| Juveniles | 74.2 | 27.2 | 67.5 | 28-183 | 63 |
| Total | 103.1 | 56.1 | 95.5 | 12-449 | 336 |

Table 1: Bone lead concentrations (µg/g dry weight) at Poundbury by sex

| | Mean | SD | Median | Range | n |
|---|---|---|---|---|---|
| 15- | 71.6 | 36.6 | 67.0 | 12-187 | 25 |
| 25- | 105.3 | 46.9 | 101.4 | 40-319 | 64 |
| 35- | 116.5 | 51.8 | 103.5 | 42-317 | 66 |
| 45- | 117.8 | 53.8 | 107.5 | 41-290 | 60 |
| 55- | 126.8 | 100.6 | 90.0 | 45-449 | 31 |
| 65+ | 104.3 | 44.9 | 92.0 | 40-215 | 21 |

Table 2: Bone lead concentrations (µg/g dry weight) at Poundbury by age

For the adults, the bone lead levels behave in accordance with a priori predictions, there is little evidence for post-mortem contamination except in the cases referred below, and they may thus be taken as a valid reflection of lead exposure. For the juveniles, however, the picture was not so reassuring. The mean lead concentration in this group (aged under 15) was much higher than present day experience would suggest was likely and there was no difference with age (table 3). It seems unlikely that all juveniles included in the study truly had the massive exposure during life which the bone concentrations suggested and some alternative explanation is much more likely.

| | Mean | SD | Median | Range | n |
|---|---|---|---|---|---|
| 0- | 71.3 | 29.6 | 69.0 | 28-119 | 8 |
| 1- | 74.3 | 28.3 | 68.0 | 43-142 | 16 |
| 5- | 75.7 | 22.6 | 72.0 | 50-132 | 20 |
| 10-15 | 73.7 | 31.4 | 63.0 | 44-183 | 19 |

Table 3: Bone lead concentrations (µg/g dry weight) in juveniles from Poundbury

## Skeletons in lead coffins

Several lead coffins were found at Poundbury and bones were taken from most for analysis. With only two exceptions, the lead levels fell outside the range of values for the main sample and in many cases the levels were extremely high (table 4) showing a considerable degree of contamination. It seemed important to determine the extent to which the lead coffins might have contaminated surrounding burials but there was almost no evidence for this. There was no systematic relationship between the lead concentration in a skeleton and its proximity to a lead coffin and only three skeletons with levels greater than 200 µg/g were found within 5 m of a lead coffin. These data give no reason to suppose that the presence of lead coffins was a factor to take into account when interpretating the results from this site.

Table 4: Lead concentrations (µg/g dry weight) in bones from lead coffins at Poundbury

| | | | |
|---|---|---|---|
| 1) | 296 | 9) | 5.467 |
| 2) | 407 | 10) | 5.649 |
| 3) | 525 | 11) | 8.469 |
| 4) | 621 | 12) | 10.052 |
| 5) | 1.109 | 13) | 13.234 |
| 6) | 1.292 | 14) | 13.287 |
| 7) | 3.241 | 15) | 14.210 |
| 8) | 4.387 | 16) | 14.362 |

## Lead exposure at different periods

The importance of determining body burdens of heavy metals (or indeed any other elements) in past populations lies in being able to demonstrate relative trends over time. There is little doubt that whatever the actual elemental concentration in the bones was during life, it will have altered after death both by substraction and addition given the fact that the bone remains chemically active. However, it is the relative not the absolute levels which are important and, provided the results obtained are biologically plausible, it is possible to compare results from different periods to determine how exposures might have varied and to speculate upon the effects which these exposures might have had on the health of the populations concerned. From the analyses which I have undertaken, it is possible to produce a preliminary comparison for some British populations (see table 5). Taking bone concentrations from a neolithic population as the standard, then it can be seen that exposure to lead increased steadily through the medieval period and that is has declined since the 18th and 19th centuries. Exposure in modern times is probably not much more than half that during the Romano-British period, mainly because there has been a dramatic decline in the amount of lead in food and drink.

Table 5: Relative lead exposure at different periods in Britain

| | |
|---|---|
| Neolithic | 1 |
| Iron age | 3.5 |
| Romano-British | 7.0 |
| Medieval | 13.0 |
| 18th/19th century | 10* |
| Present day | 4+ |

* Estimated from documentary sources.
+ Based on analyses on rib samples obtained at autopsy.

## Other metals

Although this chapter is concerned with the heavy metals, it may be appropriate briefly to touch upon some of the other biologically important metals. Strontium and selenium are dealt with extensively elsewhere in this volume and I will not add to what has been said there.

Iron: Of the other metals, iron is by far the most significant to human health and in the present day, iron deficiency anaemia is one of the most common diseases throughout the world and there is no reason to suppose that it was not also common in antiquity. Modest degrees of iron deficiency do not produce any skeletal changes except, perhaps, cribra orbitalia, although this association is by no means universally acceded to.[11] Nevertheless, iron concentrations in the skeleton are high (>100 ppm) and iron status could readily be determined from the measurement of bone levels. Some preliminary work has already been undertaken. For example, Zainio[12] was able to show the presence of iron in the bones from Anasazi Indians although none was considered to be iron deficient. Fornaciari and his colleagues[13], in a study of bones from a Punic tomb at Carthage, showed that skulls with cribra orbitalia had a lower iron content than those without and this difference was particularly obvious in females. The number of individuals in the study was small (only 12) and this limits the power of the observations but these results are most interesting and deserve to be followed up.

Zinc: Zinc is another element which is relatively abundant in the skeleton - levels range between 10-100 ppm - and this makes it a candidate for palaeopathological study. Zinc deficiency has been found to be endemic in young males in Iran and Egypt where it is associated with nutritional dwarfism and hypogonadism.[14] A study of skeletons with growth retardation might prove of interest in the future.

Copper: Copper levels in the skeleton are between 1-10 ppm and it is mentioned here only because copper deficiency causes osteoporosis in animals secondary to an impairment in collagen metabolism.[15] There has been some discussion in the medical press recently about the possible role of copper deficiency in fractures seen in infants.[16]

## Conclusion

Elemental analysis of the skeleton undoubtedly has a role to play in determining the health of past populations provided that the methodology is standardised and the possibility of post-mortem contamination can be discounted or at least counteracted. One danger in this approach, however, is to make assumptions which are not warranted in the light of present clinical knowledge. It is imperative that any hypothesis which are to be tested on past populations are based on data derived from modern populations both normal and abnormal. If modern data are not available, then any new hypothesis about trace elements and health must first be tested on the living and not the dead.

## Notes and references


1. For details of environmental mercury poisoning see Environmental Health Criteria: 1. Mercury, World Health Organisation, Geneva, 1976

2. Itai-itai disease, a form of senile osteoporosis which occurred in multiparous women in Japan, was thought by some to be caused by the use of cadmium rich water for irrigating rice. The water was taken from a river which had drained through a zinc-mining area. The disease is most likely to be a form of vitamin D deficiency, however, which may have been exacerbated by cadmium. Further information may be obtained from Cadmium and Health (ed L. Friberg, C.-G. Elinder, T. Kjellström & G.F. Nordberg), CRC Press, Boca Raton, 1985

3. In De architectura (VIII, 6), Vitruvius states that 'water is much more wholesome from earthenware than from led pipes. For it seems to be made injurious by lead because cerusse is produced by it; and this is said to be harmful to the human body'.

4. Paul of Aegina who wrote in the 7th century AD describes a colic 'having taken its rise in the country of Italy... which in many cases terminates in epilepsy, but in others in paralysis of the extremities' (De re medica, III, 43). This is a clear reference to lead poisoning with encephalopathy and peripheral neuropathy.

5. Barry PSI (1975) A comparison of concentrations of lead in human tissue. Brit J Ind Med 32: 119-139. Brätter P, Gawlik D, Lausch J, Rösick U (1977) On the distribution of trace elements in human skeletons. J Radioanal Chem 37: 393-403.

6. There is a steady increase in the concentration of deciduous teeth, for example, from first incisor to second molar. This difference may be two or three fold. Mackie AC, Stephens R, Townshend A, Waldron HA (1977) Tooth lead levels in Birmingham children. Arch Environ Health 32: 178-185.

7. See, for example, Department of the Environmental Central Directorate on Environmental Pollution: Pollution Report No 18. European Community Screening Programme for Lead - United Kingdom Results, HMSO, London, 1983.

8. Waldron HA (1981) Post-mortem absorption of lead by the skeleton. Am J Phys Anthrop 55: 395-398. But see also Lambert JB, Simpson SV, Buikstra JE, Hanson D (1983) Electron microprobe analysis of elemental distribution in excavated human femurs. Am J Phys Anthrop 62: 409-423, who find no post-mortem uptake, presumably because the soil conditions inhibited the free movement of lead.

9. Waldron HA (1983) On the post-mortem accumulation of lead by skeletal tissues. J Archaeol Sci 10: 35-40.

10. Waldron HA (1982) Lead in bones: a cautionary tale. Ecology of Disease 1: 191-196.

11. Stuart-Macadam P (1985) Porotic hyperostosis: representative of a childhood condition. Am J Phys Anthrop 66: 391-398. Kent S (1986) The influence of sedentism and aggregation on porotic hyperostosis and anaemia: a case study. Man 21: 605-636.

12. Zainio EC (1968) Elemental bone iron in the Anasazi Indians. Am J Phys Anthrop 29: 433-436.

13. Fornaciari G, Mallegni F, Bertini D, Nuti V (1983) Cribra orbitalia, and elemental bone iron in the Punics of Carthage. Ossa 8: 63-77.

14. Underwood EJ (1977) Trace Elements in Human and Animal Nutrition. 4th edition, Academic Press, London. pp 217-218.

15. Underwood (n14), pp 76-77.

16. See the discussion between Chapman and Paterson, for example. Chapman S (1987) Child abuse or copper deficiency? A radiological view. Brit Med J 294: 1370. Paterson CR (1987) Child abuse or copper deficiency? Brit Med J 295: 213-124.

# SOIL FEATURES AND HUMAN HEALTH

Iain Thornton
Applied Geochemical Research Group
Department of Geology
Imperial College of Science and Technology
London SW7 2BP, UK

## Introduction

Human health has in the past and continues to be influenced by the amounts and proportions of chemical elements to which human beings are exposed through food, drinking water and the atmosphere. Direct ingestion and inhalation of soil and dust are also sometimes important. Some of these elements are nutritionally essential, others may be toxic. Most of the essential elements may also be toxic if present in large amounts. Trace elements and metals are considered in this article. Their essentiality and toxicity has been reviewed by Underwood (1977), Crounse et al. (1983), Moynahan (1979) and many other authors. They list cobalt (Co), chromium (Cr), copper (Cu), fluorine (F), iodine (I), iron (Fe), manganese (Mn), molybdenum (Mo), selenium (Se), and zinc (Zn) as essential; nickel (Ni), tin (Sn), silicon (Si), and vanadium (V) as possibly essential, and aluminium (Al), arsenic (As), cadmium (Cd), lead (Pb), and mercury (Hg) as toxic.

Regional variations in the prevalence of human disease are recognised in many parts of the world, and both epidemiologists on the one hand and geochemists on the other have attempted to seek relationships between these geographic patterns of disease and the geochemical environment. This rapidly advancing research area of geochemistry and health is concerned primarily with the recognition and understanding of the relationship between the composition of rocks, soils, dust, airborne materials, ground and surface waters, and the health of plants, livestock, and man (Webb 1964; Thornton & Webb 1977). The geochemist is also concerned with the interaction between natural geochemical and anthropogenic sources of metals in the environment (Thornton et al. 1986), particularly where these metals have been derived from mining and smelting. Metal pollution dates back to prehistoric times, and toxicity of metals to plants, animals and man has been known for centuries (Thornton & Abrahams 1984). However, it is probable that toxic effects due to either industrial or environmental exposure were frequently not recognised nor diagnosed correctly at the time. Relatively recently it has been suggested that the fall of the Roman Empire was in part due to endemic lead poisoning brought about by the use of lead lined pots used for cooking and storing wine (Gilfillan 1968) and it has been suggested that Napoleon may have died of arsenic poisoning resulting from the use of copper arsenite as a pigment for colouring wallpaper (Jones 1982).

This paper outlines several ways in which the chemical nature of soil and related earth materials may influence today and may have influenced in the past the health and well being of the population. It is also realised that soil may be a source of contaminants that may over time affect the chemical nature of buried human remains.

## Sources of trace elements and metals in soils

The main sources of trace elements in soils are the parent materials from which they are derived. Usually this is weathered bedrock or overburden that has been transported by wind, water or glacial activity. Overburden may be local or exotic, though in Britain, transported material is mainly of local origin. Ninety-five percent of the earth's crust is made up of igneous rocks and 5% sedimentary rocks; of the latter about 80% are shales, 15% sandstones and 5% limestones (Mitchell 1964). Sedimentary rocks tend to overlie the igneous rocks from which they were derived and hence are more frequent in the surface weathering environment. The degree to which trace elements in igneous rocks become available on weathering depends on the type of minerals present. The more biologically important trace elements such as Cu, Co, Mn and Zn occur mainly in the more easily weathered materials such as hornblende and olivine (Mitchell 1974). Of the sedimentary rocks, sandstones contain minerals that weather slowly and usually contain only small amounts of trace elements. On the other hand, shales may be of organic or inorganic origin, and usually contain large amounts of trace elements (Mitchell 1964). Black shales in particular are enriched in a number of elements including Cu, Pb, Zn, Mo and Hg, sometimes at levels deleterious to plant and/or animal health.

Soils derived from these parent materials tend to reflect their chemical compositions. Thus those developed from the weathering of coarse-grained sediments such as sands and sandstones, and from acid igneous rocks such as rhyolites and granites tend to contain smaller amounts of nutritionally essential elements such as Cu and Co, than those derived from fine-grained sedimentary rocks such as clays and shales and from basic igneous rocks. Potentially toxic amounts of trace elements in soils may be derived as the result of weathering of metal-rich source rocks (Table 1). For example, some calcareous soils developed from inter-bedded shales and limestones of the Lower

Lias formation (Jurassic) in southwest England contain 20 ppm Mo or more, and are associated with molybdenosis and molybdenum-induced copper deficiency in grazing cattle.

The influence of parent materials on the total content and form of trace elements in soils is modified to varying degrees by pedogenetic processes that may lead to the mobilization and redistribution of trace elements both within the soil profile and between neighbouring soils. In the United Kingdom and similar temperate areas, most of the soils are relatively young and the parent material remains the dominant factor. Under tropical climates and on more mature land surfaces, such as those in Australia, weathering processes have been more vigorous or of much greater duration and relationships between the chemical composition of the original parent materials and the soil may be completely overridden by the mobilization and secondary distribution of chemical elements and the formation of secondary minerals.

Table 1. Trace elements and metals in soils derived from normal and geochemically anomalous parent materials in Britain

| Typical normal range in soil (ppm) | | Metal-rich soils (ppm) | Sources | Possible effects |
|---|---|---|---|---|
| As | <5-40 | up to 2500<br>up to 250 | Mineralization<br>Metamorphosed rocks around Dartmoor | Toxicity in plants and lifestock; excess in food crops |
| Cd | <1-2 | up to 30<br>up to 20 | Mineralization<br>Carboniferous black shale | Excess in food crops |
| Cr | 15-300 | up to 3500 | Ultrabasic rocks in Scotland | No known effect |
| Cu | 2-60 | up to 2000 | Mineralization | Toxicity in cereal crops |
| F | 20-500 | 1% or more | Mineralization | Fluorosis in lifestock |
| Hg | 0.008-0.19 | 1-7 | Mineralization | No known effect |
| Mo | <1-5 | 10-100 | Marine black shales of varying age | Molybdenosis or molybdenum-induced hypocuprosis in cattle |
| Ni | 2-100 | up to 8000 | Ultrabasic rocks in Scotland | Toxicity in cereal and other crops |
| Pb | 10-150 | 1% or more | Mineralization | Toxicity in lifestock; excess in foodstuffs |
| Se | <1-2 | up to 7 | Marine black shales in England and Wales | No effect |
| | | up to 500 | Namurian shales in Ireland | Chronic selenosis in horses and cattle |
| Zn | 25-200 | 1% or more | Mineralization | Toxicity in cereal crops |

The process of gleying, leaching, surface organic matter accumulation and podzolization, together with soil properties such as reaction (pH) and redox potential (Eh) may affect the distribution, the form and the mobility of trace elements in the soil. Trace elements, including Se, are often leached from the surface layers of podzols and enriched in the B horizon (Smith 1983).

Soil in mineralized areas are often enriched in the ore metals and in Britain frequently contain high concentrations of one or more of the elements Cu, Pb, Zn, Cd and As. Factors controlling metal dispersion from mineralized sources in residual soils, glacial drift, alluvial soils and organic soils have been reviewed by Hawkes & Webb (1962), Levinson (1974) and Seigel (1974). Anthropogenic sources of metals in soils include (i) industrial emissions and effluents, (ii) traffic, (iii) urban development and dumped waste materials, (iv) contaminated dusts and rainfall, (v) sewage sludge, pig slurry and composted refuse and (vi) fertilizers, soils ameliorants and pesticides.

## The soil-plant-animal (human) relationship

The availability of trace elements in the soil to plants is controlled both by their total concentration and chemical form. The chemical forms or 'species' of trace elements in soils are still poorly understood and the subject has been comprehensively reviewed by Sposito (1983) and Mattigod and Page (1983). Soil and plant factors influencing the mobilization of trace elements in soils, availability and uptake by plants and subsequent translocation within the plant have been discussed by Mitchell & Burridge (1979), West (1981) and Tinker (1981). The intake of metals by grazing animals depends on the nature of their diet and is affected by species composition of the sward and season of the year, amount of supplementary feed provided, selectivity of grazing and the availability of the element and its absorption within the alimentary tract. There are still large gaps in the knowledge of factors influencing trace metal availability to the animal, particularly in relation to different sources and forms of the metal and interactions with other dietary components.

It has also been recognised that grazing animals involuntary ingest soil along with grass. In Britain and New Zealand, cattle may ingest from 10 to over 100g soil/kg herbage DM, and sheep, which graze closer to the ground up to 300g (Thornton 1974). The amount of soil ingested will depend on the type of pasture, management factors, soil type, season and weather factors, such as rainfall intensity (Healy 1969). Soil ingestion is thought to be an important pathway of metals into animals in areas where soils are contaminated with or contain naturally high concentrations of heavy metals. Recent studies in southwest and central England, in areas where past metalliferous mining and smelting have led to widespread soil contamination with one or more of the metals Cu, Pb, Zn, Cd and As, have concluded that only a small proportion of these metals is taken up into the leafy material of pasture plants and that plant uptake is not the major pathway of these contaminants into grazing animals. Calculations based on soil, plant and faecal analysis have shown that, under the prevailing conditions, from 9 to 80% of the Pb and 34 to 90% of the As intake into cattle is due to ingested soil (Thornton & Kinniburgh 1978, Thornton & Abrahams 1981).

Pathways of metals to man through the food chain or by accidental inhalation and ingestion are also subject to many chemical and physical processes depending on the nature and source of the elements, their mobility, solubility and availability to plants and within the human gut, chemical and biochemical interactions in the soil, plant, animal and human, and interactions with other dietary constituents. It is frequently argued that in the modern world, health and disease in man can no longer be expected to reflect his local environment due to the diverse nature and sources of his diet and to his mobility. However, rural communities in the 'western' world and many people in developing countries live much 'closer to the ground' - as did our forebears, with much greater dependence on home grown crops and local water supplies than the town dweller of today.

## Trace element maps

Studies into the relationship between geochemical parameters and health are greatly assisted by knowledge of the regional distribution of chemical elements. The spatial distribution of rocks and of soils developed from different parent materials will frequently govern the levels of elements in the surface environment. When this is so, published geological maps will provide useful information on the likelihood of natural deficiency or excess. However, the scales of such maps are often too large to pinpoint localities in which disease may occur or else the units used for mapping may be too general to show changes in lithological features. Similarly the average tabulated abundances of elements in particular rock types (Clarke values) do not always represent specific local situations. For example some Scottish granites exceed 36ppm U compared with the accepted Clarke value of 4.8 ppm (Plant & Stevenson 1985).

Maps based on systematic sampling and analysis of soil and vegetation would provide information closer to man's food chain, though this is often not possible because of cost and time.

The requirement for systematic data on trace element levels over the United Kingdom has been met in part by geochemical reconnaissance surveys, carried out by the Applied Geochemistry Research Group (AGRG) of Imperial College and the British Geological Survey (previously known as the Institute of Geological Sciences). These surveys are based on the collection and analysis of stream sediment samples, on the premise that each sample of sediment approximates to a composite sample of the erosion products of rock, overburden and soil upstream from the sampling point, and hence represents the average concentration of elements in the sampling area. The data obtained from these surveys are available in published atlases and in machine readable form.

An example of one of the maps from the Wolfson Geochemical Atlas of England and Wales, compiled by AGRG (Webb et al. 1978), is shown in Plate 1. This map for lead is based on the analysis of 50,000 stream sediment samples taken over 64,000 square miles throughout a 10-week period in 1969. The Atlas, published at a scale of 1:2 million contains computer-smoothed maps for 21 elements. Geochemical atlases are now available for parts of Scotland, Northern Ireland, West Germany, parts of the United States, and some Scandinavian countries. Surveys are in progress in many other countries.

## Applications to health

The majority of the research into the applications of geochemical maps to health have been concerned with crop and animal nutrition and with pollution studies. These have been made successively reviewed by Thornton & Webb (1979), Thornton & Plant (1980) and Thornton (1985), and some documented in the proceedings of the conferences on 'Trace Substances in Environmental Health' held annually since 1967 at the University of Missouri (Hemphill 1967-87) and in the Proceedings of the First International Symposium on Geochemistry and Health (Thornton 1986). The application of geochemical maps to human health has received less attention, probably due to the limited success in relating causally the geochemical environment and human disease (Thornton 1987). Obvious exceptions are (a) the well documented relationship between the fluorine content of waters and the incidence of human fluorosis (and inverse relationship with dental caries); (b) the relationship between the prevalence of endemic goitre and iodine levels in water supplies and food; and (c) the prevalence of Se-responsive diseases in areas of low Se status in parts of the Peoples Republic of China.

Two studies have been undertaken into the possible effects of human exposure to metals in "geochemical hotspot" situations highlighted by the Wolfson Atlas maps. They are both related to widespread metal contamination of the surface environment due to historical mining and smelting activities.

The map for lead (Plate 1) shows anomalously high levels in several discrete areas, totalling around 4,000 $km^2$. In an area of past lead-zinc mining in Derbyshire, a collaborative study by the Paediatric Unit, St. Mary's Hospital Medical School and AGRG, showed that when households were grouped according to the lead content of the soil in the gardens, the amounts of lead in the blood and hair of the 2 to 4-year old children increased with that in the soil and house dust (Table 2). However, none of the lead values in children were high enough to be considered hazardous at that time, even though the amounts in the soil and dust were extremely high, peaking at 2.8 and 2.5% lead, respectively. The children's lead status does, however, reflect that of the environment and, in the absence of evidence linking the human burden with lead in foodstuffs or water supplies, it was suggested that the major pathway of lead in the children was through the inhalation of and the involuntary ingestion of dust particles (Barltrop et al. 1975). A comprehensive environmental health study was conducted around the village of Shipham in south-west England shown by the map for cadmium (not illustrated). Zinc, mined as the ore smithsonite ($ZnCo_3$), containing appreciable amounts of Cd, had been mined over the period 1700 to 1850. Collaborative investigations involving local and national government, Westminster Medical School and AGRG

Table 2. Lead content of garden soil, house dust, blood and hair

| Soil (ppm) | House dust (ppm) | Blood (µg/100 ml) Child | Blood (µg/100 ml) Mother | Hair (ppm) Child |
|---|---|---|---|---|
| 130-900 | 190-2,450 | 15-33 | 9-40 | 2-36 |
| 420 (24) | 531 (20) | 21 (29) | 14 (25) | 8 (29) |
| 1,050-9100 | 390-25,000 | 13-45 | 11-44 | 2-62 |
| 3,390 (40) | 1,564 (36) | 24 (43) | 19 (41) | 11 (43) |
| 10,000-28,000 | 1,000-5,100 | 21-43 | 13-19 | 10-39 |
| 13,969 (8) | 2,582 (8) | 29 (10) | 15 (8) | 29 (10) |

Range and geometric mean; number of samples in parentheses.

showed (a) over 60% of household garden soils with more than 60 ppm Cd, with Zn:Cd ratios averaging 90:1; (b) household dust averaging 26 ppm Cd and 2,300 ppm Zn; (c) an increased dietary Cd intake compared to the U.K. national norm, though rarely exceeding the World Health Organisaion's provisional tolerable weekly intake of 450 to 500 µg Cd. Health inventories and biochemical tests on 548 residents of Shipham and on 543 control subjects from a nearby uncontaminated village showed only slight differences attributable to cadmium. In this unique situation it is possible that the large amounts of zinc and cadmium present in the environment afford some degree of protection against the possible adverse effects of cadmium.

## Urban geochemistry

Methods developed by the geochemist have recently been applied to further the understanding of the chemical nature of the urban environment.

In 1981 a national survey of metals in dusts and soils was commissioned by the U.K. Department of the Environment. The prime aim of this study was to establish the usual ranges of metal concentration to which the population may be exposed in the home and urban environment. Over 5,000 households were examined covering 53 city boroughs, towns and villages in England, Scotland and Wales. The results showed the presence of elevated concentrations of Pb, Cd, Cu and Zn, which ranged widely in both soils and floordusts (Thornton et al. 1968; Culbard et al. 1987). The geometric mean values for Pb in housedusts and garden soils for all localities (excluding old mining areas) were 561 ppm and 266 ppm respectively. Over 10% of floordusts exceeded 2,000 ppm Pb. In the Derbyshire lead mining area, housedusts averaged 1,870 ppm Pb and garden soils 5,600 ppm Pb, clearly illustrating the 'geochemical hotspot' effect. This may be compared with a mean value of 205 ppm Pb in floordust for a small residential town in Scotland.

Current studies are focussed on the importance of lead exposure through transfer of lead-rich dust by hand-to-mouth activity in young children. This pathway must surely have been of even greater significance when human hygiene was of less importance.

## Conclusions

Concentrations of trace elements and heavy metals in soils may vary appreciably from one locality to another due to the natural geochemical composition of parent materials including underlying bedrock. Contamination from mining and metal processing and from other anthropogenic sources gives rise to very large concentrations of metals in surface soils in specific areas. Only a limited number of causal relationships have been established between the distribution of chemical elements in the environment and human disease. These are due to deficiencies of F, $I_2$ and Se and excesses of F. Excesses of Pb are of possible concern to the health of young children. Man's trace element status and exposure to metals is commonly reflected in the composition of tissues and body fluids. Possible fingerprints of environmental history relating to diet and exposure may be shown by amounts of Se in hair, and F and Pb in teeth and bone. Geochemical maps provide a useful means of delineating areas of trace element deficiency and metal excess and are likely to have direct application to studies into man's environment in antiquity.

## References


Barltrop D, Strehlow CD, Thornton I & Webb JS (1975) Absorption of lead from dust and soil. Postgrad Med J 52: 801-804

Barltrop D & Strehlow CD (1982) Cadmium and Health in Shipham. Lancet: 1982

Crounse RG, Porries WJ, Bray JT & Maugher RL (1983) Geochemistry in man: health and disease. In: Thornton I (ed) Applied Environmental Geochemistry. Academic Press London. pp 267-333

Culbard E, Thornton I, Watt J, Wheatly M, Moorcroft S & Thompson M (1987) Metal contamination in British urban soils and dusts. J Environ Quality. In press

Gilfillin SC (1965) Lead poisoning and the fall of Rome. J Occup Med 7: 53-60

Hawkes HE & Webb JS (1962) Geochemistry and Mineral Exploration. Harper & Row New York Evaston

Healy WB (1969) The influence of soil type on ingestion of soil by grazing animals. 9th Int Cong Soil Sci Trans 3: 437-445

Hemphill DD (ed) (1967-1986) Trace Substances in Environmental Health 1-XX, University of Missouri, Columbia, Missouri

Jones D (1982) The singular case of Napoleon's wallpaper. New Scientist 96: 101-104

Levinson AA (1974) Introduction to Exploration Geochemistry. Applied Publishing Ltd Calgary

Mattigod SV & Page AL (1983) Assessment of metal pollution in soils. In: Thornton I (ed) Applied Environmental Geochemistry. Academic Press London. pp 355-394

Mitchell RL (1964) Trace Elements in Soils. In: Bear FE (ed) Chemistry of the Soil 2nd ed. Reinhold Publ Co New York. pp 320-368

Mitchell RL (1974) Trace elements problems in Scottish soils. Neth J Agric Sci 22: 295-304

Mitchell RL & Burridge JC (1979) Trace elements in soils and crops. Philos Trans R Soc London B288: 15-24

Moynahan EJ (1979) Trace elements in man. Philos Trans R Soc London B288: 65-79

Plant J & Stevenson AG (1985) Regional geochemistry and it's role in epidemiological studies. In: Mills CJ et al (ed) Trace Elements in Man and Animals. Commonwealth Agricultural Bureau, Farnham Royal. pp 900-906

Seigel FR (1974) Applied Geochemistry. Wiley New York

Smith CA (1983) The Distribution of Selenium in some Soils developed on Silurian, Carboniferous and Cretaceous Systems in England and Wales. PhD Thesis, University of London

Sposito G (1983) The chemical forms of trace elements in soils. In: Thornton I (ed) Applied Environmental Geochemistry. Academic Press London. pp 123-170

Sposito G (1983) The chemical forms of trace elements in soils. In: Thornton I (ed) Applied Environmental Geochemistry. Academic Press London. pp 123-170

Thornton I (1974) Biogeochemical and soil ingestion studies in relation to trace element nutrition of lifestock. In: Hoeckstra WG et al (ed) Trace Element Metabolism in Animals 2. University Press Baltimore. pp 451-454

Thornton I & Webb JS (1979) Geochemistry and health in the U.K. Philos Trans R Soc London B288: 151-168

Thornton I & Kinniburgh D (1978) Intake of lead, copper and zinc by cattle from soil and pasture. In: Kirchgessner M (ed) Trace Element Metabolism in Man and Animals Freising

Thornton I & Abrahams P (1984) Historical records of metal pollution in the environment. In: Nriagu JO (ed) Changing Metal Cycles and Human Health. Springer-Verlag Berlin Heidelberg New York. pp 7-25

Thornton I (1985) Environmental geochemistry and health in the U.K. In: Lag J (ed) Geomedical Research in Relation to Geochemical Registration. Universitetforlagetm Oslo. pp 125-136

Thornton I, Culbard E, Moorcroft S, Watt J, Wheatley M, Thompson M & Thomas JFA (1985) Metals in urban dusts and soils. Environmental Technology Letters 6: 137-144

Thornton I (ed) (1986) Proceedings of the First International Symposium on Geochemistry and Health. Science Reviews Ltd London

Thornton I, Abrahams PW, Culbard E, Rother JAP & Olson BH (1986) The interactions between geochemical and pollutant metal sources in the environment. Implications for the community. In: Thornton I & Howarth RJ (eds) Geochemistry in the 1980's. Graham & Trotman Ltd London

Thornton I (1987) Mapping of trace elements in relation to human disease. Clin Nutr 6: 97-104

Tinker PB (1981) Levels, distributions and chemical forms of trace elements in food plants. Philos Trans R Soc London B294: 41-55

Underwood BJ (1977) Trace Elements in Human and Animal Nutrition. 4th ed. Academic Press New York

Webb JS (1964) Geochemistry and life. New Scientist 23: 504-507

Webb JS, Thornton I, Thompson M, Howarth RJ & Lowenstein PL (1978) The Wolfson Geochemical Atlas of England and Wales. Oxford University Press Oxford

West TS (1981) Soil as a source of trace elements. Philos Trans R Soc London B294: 19-39

# THE BIOAVAILABILITY OF TRACE ELEMENTS AND AGE-SPECIFIC TRACE ELEMENT METABOLISM

H. Zumkley, C. Spieker
Medizinische Universitäts-Poliklinik
Albert-Schweitzer-Str. 33
4400 Münster, FRG

Abstract

Expanding knowledge upon the role of trace elements in biological systems has directed the attention more and more towards the implications of trace elements in human health. In this context, the bioavailability plays an important role. Unfortunately our knowledge upon the bioavailability of trace elements is still insufficient. A number of diseases and clinical symptoms associated with deficiency or an increase of essential respectively non-essential elements is known, but further intensive investigations are still required. Our knowledge concerning age-specific abnormalities in trace element metabolism is even very low. Of some essential and toxic elements, only age-specific alterations of trace element concentrations are known.

## Introduction

During the last years there has been a growing interest in the metabolism of essential and non-essential metals. The expanding knowledge herewith associated upon the role of trace elements in biological systems has directed the attention more and more towards the implications of trace elements in human health.

Considering the multiplicity of the roles in which the essential trace elements are involved, there must exist mechanisms to provide adequate supply for cellular requirements to prevent disturbances of the metabolism. On the other hand protective mechanisms must also exist to prevent an uncontrolled increase of the essential as well as of the non-essential, possibly toxic elements. In this context, the bioavailability plays an important role. We understand by the term bioavailability the extent and the rapidity by which essential ressources such as trace elements reach the place (cells) of their biological function.

Factors involved in the bioavailability of trace elements are:
1. intake and absorption
2. excretion
3. carrier mechanism
4. physiological state of the subject
5. intake of drugs.

## 1. Intake and absorption

Factors influencing absorption and intake of trace elements are their concentration in the soil, including geographical incidents and pH, nutrition, dietary constituents, regulation of absorption, the site of absorption and the interaction with other metals.

A characteristic disease probably caused by selenium deficiency is "Keshan Disease". This disease was observed in an area of China with a low concentration of selenium in the soils. Predominantly children and women were affected by this disease. The clinical signs are a cardiomyopathy with arrythmias, cardial insufficiency and finally death. This was observed in 50% of the patients. Selenium concentrations in blood and hair of the patients were decreased. After supplementation of selenium, the disease disappeared in those districts of China (Chang 1985).

Further areas containing decreased selenium concentrations were found in other countries as New Zealand and the United States.

Another defect also observed in China is Kaschin-Beck-disease. The patients suffer from a dysfunction of the ossification of bone, osteochondrosis and growth retardation. In some cases, the symptoms disappeared after selenium supplementation (Lombeck 1985).

Besides the Keshan-disease a number of other diseases associated with selenium deficiency is known (tab. 1). In some cases the disease was due to selenium deficiency, whereas in other cases the selenium-deficiency was due to the treatment of the disease.

If the serum selenium concentrations are discussed particularly with respect for childhood, age-depending changes in serum concentration, in blood and tissue have to be considered. The highest blood selenium concentrations were found at birth. After birth the selenium levels decreased in the different tissues as was demonstrated by Lombeck

(Lombeck 1985). Furthermore, very high selenium levels could be found in premature children and in foetal tissues.

Table 1: Diseases associated with selenium deficiency.

1. Diminished intake and/or absorption
2. Total parenteral nutrition
3. Keshan disease
4. Phenylketonuria
5. Maple-sirup-disease
6. Pancreatitis
7. Diseases with no regular Se-deficiency:
   Myocardial infarction, Arteriosclerosis,
   Alcoholism, liver diseases,
   Multiple sclerosis, neuronale Lipofuscinosis,
   Diabetic retinopathy, Katarakt,
   Arthritis, Legionnairise disease,
   Cancer.

## 2. Excretion

Five ways of excretion for trace elements are known. Fecal excretion is dependent particularly from the supply. Urinary excretion of trace elements is predominantly dependent from intake and absorption of trace elements. Biliary and sweat excretion of trace elements are very low. An important factor for the bioavailability of trace elements for children is the excretion of trace elements in the mother's milk.

The concentration of trace elements in cow's milk was found to differ partly from that of human milk. Thus, the copper concentration of cow's milk (0.06 - 0.09 µg/ml) il lower than in human milk (0.24 - 0.05 µg/ml) whereas the concentration of zinc is higher in cow's milk (3.23 - 5.15 µg/ml) than in human milk (1.16 - 3.83 µg/ml). Furthermore, cow's milk contains about 4 - 5 times more calcium and magnesium. The concentration of iron in cow's milk (0.40 - 0.59 µg/ml) is similiar to that of human milk (0.020 - 0.69 µg/ml) (Fransson et al. 1984).

The bioavailability of the trace elements in milk is not only dependent on the concentration but also on the binding to different protein fractions. Thus, the casein fraction in cow's milk contains a large proportion of the total amount of the elements as iron, copper, zinc, calcium and magnesium, whereas human casein only binds minor amounts (Fransson et al. 1984).

The manganese concentration in cow's and human milk shows a different binding to casein, lipid fraction and whey. The different binding is probably important for the bioavailability of manganese (Lönnerdal et al. 1985). Furthermore it has been found that the manganese concentrations of some infant formulas exceed the concentration of manganese in human milk. This high concentration of manganese may be dangerous to the infants.

In animal studies it could be shown that the bioavailability of zinc was 28% from human milk, 24% from whey-adjusted cow's milk formula, 15% from cow's milk and 10% from soy formula (Sandström et al. 1983).

In all studies it could be demonstrated that lead concentrations in human milk were significantly dependent from the mother's diet (fig. 1). The data demonstrate that the

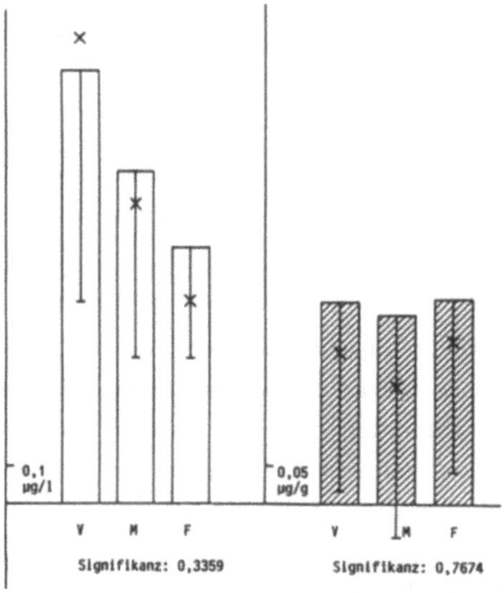


Figure 1. Influence of the diet on lead concentrations in mother's milk and hair. V = vegetable diet, M = mixed diet, F = meat enriched diet.

highest lead concentration in human milk was found in patients with predominantly vegetal diet compared to women with a diary respectively with a predominantly meat enriched diet. Furthermore, a small difference was found between the concentration of lead in human milk between winter and summer seasons. In the summer season, the lead concentration was a little lower than in winter.

Furthermore, we devided our sample of lactating women in groups living in houses built before 1920, between 1920 and 1939 as well as in women living in houses built between 1940 - 1959 and after 1960. In these studies it could be shown that lead concentrations were higher in the atmosphere of older houses compared with newer ones, thus causing higher blood lead concentrations in females living in older houses.

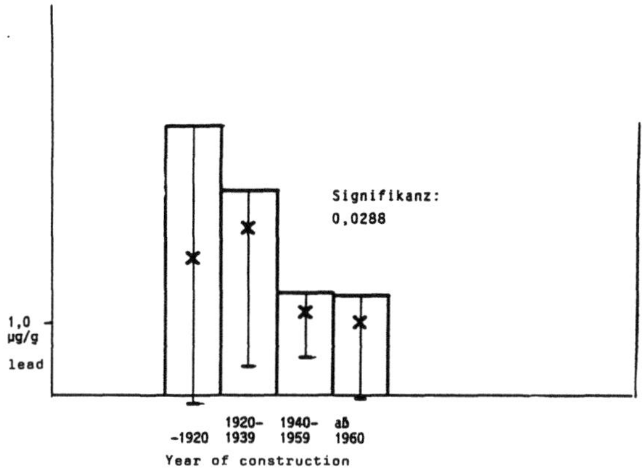


Figure. 2. Correlation between the lead concentrations in hair and the year of construction of the dwelling house.

3. Carrier mechanism

Besides the protein binding of trace elements metallothionein is one of the most important factors for the homeostasis of trace elements. Metallothionein represents a class of low molecular weight metal-and sulfur-rich proteins. They occur in many different species and tissues. The physiological roles of metallothionein are not well understood yet for all functions. However, a number of functional characteristics are already known (Vasak 1985). They include: Metal dispensing in the synthesis of metalloproteins, metal storage, metal detoxification, metal homeostasis (resorption, excretion, redistribution) and finally scavenging of free oxygen radicals.

It has to be pointed out that metallothionein seems to be the most important factor in trace element homeostasis. It is assumed that the intestinal cells must be capable of monitoring the status of the essential and non-essential trace elements. The intestinal cells give information for the synthesis of metallothionein to coordinate the flux of trace elements through the cells. Furthermore, it is believed that the absorption of toxic elements such as cadmium can be influenced by the new synthesis of metallothionein for detoxification of the toxic elements. Further intensive investigations on this field of trace elements will give us new aspects on the bioavailability of the metals.

## Drugs

The bioavailability of trace elements can often be influenced by drugs. Thus, drugs containing trace elements were often used in medicine since more than 4.000 years (Zippe 1967). We analyzed the aluminium concentrations in patients without renal insufficiency in the brain-bone-liver-gastricmucosemembrane and in the intestinal mucosa tissue after treatment with aluminium containing antacids. We found an increase of the aluminium concentrations in all tissues analyzed.

## References


Chang SL (1985) Selenium affects heart death rates in China and USA. Trace Subst Environ Health (Hemphill DP ed) Vol XIX, p264.

Forbes GB, Reina JC (1972) Effect of age on gastrointestinal absorption (Fe, Sr, Pb) in the rat. J Nutr 102: 647.

Fransson GB, Lönnerdal B (1984) Iron, copper, zinc, calcium and magnesium in human milk fat. Am J Clin Nutr 39: 185.

Lönnerdal B, Bell JG, Keen CL (1985) Copper absorption from human milk, cow's milk and infant formulas using a suckling rat model. Am J Clin Nutr 42: 836.

Lombeck I, Kasparek K, Bonnermann B, Feindegen LE, Bremmer HJ (1978) Selenium content of human milk, cow's milk and cow's milk infant formula. Europ J Pediatr 129: 139.

Lombeck I, Menzel H, Kasparek K (1985) Selenmangel im Kindesalter. In: Gladtke E, Heimann G, Lombeck I, Eckert E (eds) Spurenelemente. Thieme Vlg, Stuttgart, New York. p 73.

Sandström B, Cederblad A, Lönnerdal B (1983) Zinc absorption from human milk, cow's milk and infant formulas. Am J Dis Child 137: 726.

Vasak M (1985) Metallothionein and its significance in the homeostasis of trace elements. In: Gladtke E, Heimann G, Lombeck I, Eckert I (eds) Spurenelemente. Thime Vlg, Stuttgart, New York. p23.

Zippe SXM (1967) Geschichte der Metalle. Sändig Vlg, Wiesbaden.

Zumkley H, Bertram HP, Lison A, Zidek W, Thiem P (1984) Influence of smoking habits on cadmium concentrations in hypertensives. Trace Elements in Medicine 2:91.

Zumkley H, Richter KD, Bertram HP, Schmidt PF, Losse H, Spieker C, Peitz U, Zidek W (1985a) Influence of low and high cadmium diet in normotensive and hypertensive SH-rats. Klin Wschr 63 (Suppl.): 70.

Zumkley H, Bertram HP, Zidek W, Losse H, Spieker C (1985b) Interaction between cadmium and calcium in normotensive and hypertensive (SH) rats. Trace Subst Environ Health (Hemphill DP ed) Vol XIX, p238.

# THE DIAGNOSIS OF DISEASE BY ELEMENT ANALYSIS

O. Oster
Department of Clinical Chemistry
University Hospital of the University of Mainz
Langenbeckstr.
6500 Mainz, FRG

The elemental composition of the human body images in principle the elements of the periodic system. In the literature, about 80 elements are reported to occur in the human body and the number is expected to increase by the introduction of methods more sensitive in detection. The elemental composition in reference man as published (1) is summarized in tab. 1. From the major elements in reference man, the determination of the elements nitrogen, calcium, phosphorus, potassium, sodium, chlorine and magnesium are of importance for the diagnosis of diseases. The importance of the element determination for the diagnosis of diseases is also documented by the numbers of element determinations done daily in a hospital: In an university hospital with 1600 beds, about 1700 - 2100 elemental analyses in body fluids are done per day. The statistic is headed by the major elements sodium, potassium, calcium, chlorine, magnesium and phosphorus. The element determination is not only done in the routine laboratory but is also an integral part of the emergency laboratory. In the routine laboratory, about 14-17% of the total analyses per day are done as element analysis while in the emergency laboratory, the element determinations per total analysis is about 20-25% (2). The dominance of major elements in the statistic does not imply that the determination of essential trace elements or toxic elements is less important for the diagnosis of diseases. In case of acute metal or metalloid intoxications the determination of the element in question may be necessary within the emergency laboratory. An early diagnosis of metal and metalloid intoxications may be life saving because of the possibility of an early onset of a specific therapy (3,4,5,6).

From our experience with regard to the diagnosis of diseases by element analysis of body fluids the following knowledge should be available:

1. A general knowledge about the amount and organ distribution of the elements and especially of the biochemical and physiological function of the element. The elements

and their biochemical function are presented in a short summary in tab. 1. However, it is not intended to give a comprehensive survey but to outline subjects one has to be aware of.

Tab. 1: Elements occuring in man and their biological function. The elemental composition with regard to the major elements (O, C, H, N, Ca, P, S, K, Na, Cl, Si) is taken from Reference Man (cf ref. 1).

| Element | Total amount of body weight (g) | (%) | Biological function |
|---|---|---|---|
| O | 43000 | 61.0 | Present in $H_2O$ and all organic matter (sugars , amino acids, lipds, nucleic acids) |
| C | 16000 | 23.0 | Present in the skeleton and all organic matter |
| H | 7000 | 10.0 | Present in $H_2O$ and all organic matter |
| N | 1800 | 2.6 | Present in amino acids, nucleic acids and most of the organic matter |
| Ca | 1200 | 1.4 | Most of the calcium is present in the skeleton as calcium phosphate (99%). 10g of the total amount of calcium is found in soft tissues and the extracellular fluid. Ca is necessary for blood coagulation and the exciteability of peripheral nerves and the muscle. Ca is involved also in muscular contractility. Regulation of plasma Ca by parathormone and calcitonin. |
| P | 580 | 0.8 | Most of the P in the body is present together with Ca in the skeleton. P in its ionic form as phosphate is present in the blood. The phosphate is bound in energentic forms to organic matter in lipids, sugars, in nucleic acids and is involved in the energy transfer within the cell. |
| S | 140 | 0.2 | Present in the essential amino acids cystein and methionin. |
| K | 137 | 0.2 | K is the main intracellular ion. Concerning the equilibrium, the hormonal control by aldosterone is involved. K and Na are responsible for the maintenance of membrane potentials. Essential for the exciteability of nerves. |
| Na | 70-95 | 0.14 | Na is the main extracellular ion. In the extracellular fluid, the main function is the maintenance of the osmotic equilibrium. The concentration in plasma is hormonally controlled by aldosterone. |
| Cl | 84-94 | 0.12 | Chloride is the counter ion of sodium and follows the sodium changes nearly passive. In so far it participates like sodium in the maintenance of the osmotic equilibrium in the extracellular fluid. |
| Mg | 19 | 0.03 | Magnesium is cofactor of many enzymes, especially of enzymes of the energy transfer system using ATP. In the cell, ATP is complexed with Mg. Mg is also necessary for the function of the ribosomes. In the muscle, Mg is involved in the contractility and the neuromuscular exciteability. |

Tab. 1, cont.

| Element | Total amount of body weight (g) | (%) | Biological function |
|---|---|---|---|
| Si | 18 | 0.03 | Silicon is an essential element of so far unknown function. It is discussed to be important for the connective tissue. |
| Fe | 3.5-5 | | Essential constituent of hemoglobin which is responsible for the oxygen transport. Iron containing enzymes are involved in the electron transport. Iron is also part of myoglobin, transferrin and ferritin. |
| Zn | 2-3 | | Zinc is cofactor of many enzymes, e.g. 5'-nucleotidase, angiotensin converting enzyme, carbonic anhydratase, LDH, $\delta$ -amino-levulinicacid dehydratase, alkaline phosphatase, prostatic acid phosphatase, dismutase, $\alpha$-D-mannosidase, carboxypeptidase A and B, collagenase, glyoxalase, most probably in DNA and RNA polymerase ect. |
| Cu | 0.080 | | Copper occurs in proteins which are linked to the electron transport chain, e.g. cytochrome oxidase. Other copper containing enzymes are ascorbic acid oxidase, lysine oxadase, superoxide dismutase, ceruloplasmin. |
| Mn | 0.012 | | Manganese is present in the mitochondrial superoxide dismutase and in glycosyl transferases. |
| J | 0.013 | | Iodine is present in the thyroid hormones thyroxin and triiodothyronine. |
| Co | 0.002 | | Cobalt is part of the vitamin B-12. |
| Mo | 0.010 | | Mo is present in the enzymes xanthine-, aldehyde- and sulfite-oxidases. |
| Se | 0.006-0.010 | | Glutathione peroxidase, an enzyme involved in the inhibition and protection of lipidperoxidation, contains four atoms selenium as seleno cysteine. |
| Cr | | | The essentiality of chromium for animals is suspected from the influence of chromium on the glucose metabolism. |

2. Since the elements are taken up with the food, knowledge on the actual amount of intake is necessary. The US National Research Council has published Recommended Daily Dietary Allowances (RDA) (7) for the elements, including trace elements. They are summarized in tabs. 2-4. The Research Council stresses "that the allowances are intended to provide for individual variations among most normal persons as they live in the USA under usual environmental stresses"(7). The RDA are not accepted without criticism. In Germany, similar recommendations have been made by the Deutsche Gesellschaft für Ernährung, which are generally a little lower (8). But in principle the RDA serve as good guidelines for the daily intake of elements perhaps with one exception: The recommended daily chromium intake for infants aged 0-6 months. If the trace element intake provided by human milk should serve as a standard for infant

nutrition and as basis for establishing nutrient requirements, it can be concluded from most recent results that the RDA for chromium with regard to infants are too high. The chromium concentration for mother's milk is reported with 0.14-0.43 ng/ml (9,10, 11).

Tab. 2: Recommended Daily Dietary Allowances for Cu, Mn, F, Cr and Se (7)

| Subjects | Age (years) | Cu (mg) | Mn (mg) | F (mg) | Cr (mg) | Se (mg) |
|---|---|---|---|---|---|---|
| Infants | 0-0.5 | 0.5-0.7 | 0.5-0.7 | 0.1-0.5 | 0.01-0.04 | 0.01-0.04 |
| | 0.5-1 | 0.7-1.0 | 0.7-1.0 | 0.2-1.0 | 0.02-0.06 | 0.04-0.08 |
| Children | 1-3 | 1.0-1.5 | 1.0-1.5 | 0.5-1.5 | 0.02-0.08 | 0.02-0.08 |
| | 4-6 | 1.5-2.0 | 1.5-2.0 | 1.5-2.5 | 0.03-0.12 | 0.03-0.12 |
| | 7-10 | 2.0-2.5 | 2.0-3.0 | 1.5-2.5 | 0.05-0.20 | 0.05-0.20 |
| | 11+ | 2.0-3.0 | 2.5-5.0 | 1.5-2.5 | 0.05-0.20 | 0.05-0.20 |
| Adults | | 2.0-3.0 | 2.5-5.0 | 1.5-4.0 | 0.05-0.2 | 0.05-0.2 |

Tab. 3: Food and Nutrition Board, National Academy of Sciences National Research Council Recommended Daily Dietary Allowances, revised 1980 (7)

| Subjects | Age (years) | Weight (kg) | Height (cm) | Protein (g) | Cu (mg) | P (mg) | Mg (mg) | Fe (mg) | Zn (mg) | J (mg) |
|---|---|---|---|---|---|---|---|---|---|---|
| Infants | 0-0.5 | 6 | 60 | kgx2.2 | 360 | 240 | 50 | 10 | 3 | 40 |
| | 0.5-1 | 9 | 71 | kgx2.0 | 540 | 360 | 70 | 15 | 5 | 50 |
| Children | 1-3 | 15 | 90 | 23 | 800 | 800 | 150 | 15 | 10 | 70 |
| | 4-6 | 20 | 112 | 30 | 800 | 800 | 200 | 10 | 10 | 90 |
| | 7-10 | 28 | 132 | 34 | 800 | 800 | 250 | 10 | 10 | 120 |
| Males | 11-14 | 45 | 157 | 45 | 1200 | 1200 | 350 | 18 | 15 | 150 |
| | 15-18 | 66 | 176 | 56 | 1200 | 1200 | 400 | 18 | 15 | 150 |
| | 19-22 | 70 | 177 | 56 | 800 | 800 | 350 | 10 | 15 | 150 |
| | 23-50 | 70 | 178 | 56 | 800 | 800 | 350 | 10 | 15 | 150 |
| | 51+ | 70 | 178 | 56 | 800 | 800 | 350 | 10 | 15 | 150 |
| Females | 11-14 | 46 | 157 | 46 | 1200 | 1200 | 300 | 18 | 15 | 150 |
| | 15-18 | 55 | 163 | 46 | 1200 | 1200 | 300 | 18 | 15 | 150 |
| | 19-22 | 55 | 163 | 44 | 800 | 800 | 300 | 18 | 15 | 150 |
| | 23-50 | 55 | 163 | 44 | 800 | 800 | 300 | 18 | 15 | 150 |
| | 51+ | 55 | 163 | 44 | 800 | 800 | 300 | 10 | 15 | 150 |
| Pregnant | | | | +30 | +400 | +400 | +150 | * | +5 | +25 |
| Lactating | | | | +20 | +400 | +400 | +150 | * | +10 | +50 |

* The increased requirement during pregnancy can neither be met by the iron content of habitual American diets nor by the existing iron stores of many women; therefore the use of 50-60mg of supplemental iron is recommended. Iron needs during lactation are not substantially different from those of nonpregnant women, but continued supplementation of the mother for 2-3 months after parturition is advisable in order to replenish stores depleted by pregnancy.

Tab. 4: Recommended Daily Dietary Allowances for electrolytes (7)

| Subjects | Age (years) | Sodium (mg) | Potassium (mg) | Chlorine (mg) |
|---|---|---|---|---|
| Infants | 0-0.5 | 115- 350 | 350- 925 | 275- 700 |
| | 0.5-1 | 250- 750 | 425-1275 | 400-1200 |
| Children | 1-3 | 325- 975 | 550-1650 | 500-1500 |
| and | 4-6 | 450-1350 | 775-2325 | 700-2100 |
| Adolescents | 7-10 | 600-1800 | 1000-3000 | 925-2775 |
| | 11+ | 900-2700 | 1525-4575 | 1400-4200 |
| Adults | | 1100-3300 | 1875-5625 | 1700-5100 |

3. For the diagnosis of diseases by element analysis additionally to the daily intake the daily losses in healthy conditions should be known. Tab. 5 summarizes the daily urinary and fecal losses of healthy populations. The daily urinary losses can be calculated also on mg creatinine or as clearance in ml plasma to be freed per minute by an element. Depending on the element, the urinary losses can be used to explain deficiency symptoms, pathophysiological symptoms and intoxications. Therefore, also the toxic elements are included in tab. 5.

Tab. 5: Urinary and fecal excretion of elements per die

| Element | Amount of element in urine | Amount of element in feces |
|---|---|---|
| Nitrogen | 7 - 25 g | 0.8 - 1.5 g |
| Sodium | 50 - 200 mmol | 0.4 - 0.53 mmol |
| Potassium | 30 - 100 mmol | 10.0 - 23.0 mmol |
| Calcium | 3 - 8.5 mmol | 9.0 - 1.8 mmol |
| Magnesium | 2.5 - 9 mmol | 2.7 - 9.0 mmol |
| Chlorine | 90 - 250 mmol | 0.4 - 3.2 mmol |
| Phosphorus | 0.8 - 2.0 g | 0.24- 1.7 g as P |
| Iron | 200 - 300 µg | 8 - 20 mg |
| Copper | 20 - 60 µg | 0.6 - 3.0 mg |
| Zinc | 300 - 800 µg | 5 - 15 mg |
| Lead | 40 µg | |
| Mercury | 8 µg | |
| Arsenic | 2-25 µg | |
| Cadmium | 8 µg | |
| Thallium | 10 µg | |
| Selenium | 13-21 µg | |
| Fluorine | 0.20-4.2 µg | |
| Iodine | 170 µg | 20 µg |

4. For the diagnosis of diseases by element analysis the knowledge of reference values in plasma, serum und whole blood is required. Most of the diagnoses of diseases is done by serum analysis. For some elements e.g. the toxic elements (Pb, Cd, As and in some cases also for Hg) the determination is done in whole blood because the whole blood element concentration correlates better with the degree of intoxication. In many disease states the simultaneous element determination in plasma and urine is necessary to obtain suggestions about the balance. (Tab. 6 summarizes reference values for the elements in serum, whole blood and hair (12)). The reported reference values are established for a population living in Mainz, Germany.

Tab. 6: Reference values for serum, whole blood and hair (12)

| Element | Serum | Whole blood | Hair |
|---|---|---|---|
| Na | 135 - 145 mmol/l | | |
| K | 3.5 - 4.8 mmol/l | | |
| Ca | 2.25 - 4.5 mmol/l | | |
| Mg | 0.65 - 1.1 mmol/l | | |
| Fe | 0.6 - 1.6 mg/l | | |
| Cu | 0.6 - 1.3 mg/l | | |
| Zn | 0.8 - 1.2 mg/l | | |

Tab. 6, cont.

| Element | Serum | Whole blood | Hair |
|---|---|---|---|
| Se | 53-105 µg/l | 65-130 µg/l | 150-600 ng/g |
| Cr | <1 µg/l | | |
| Ni | <2 µg/l | | |
| Co | ≈2 µg/l | | |
| Al | <4 µg/l | | |
| Hg | <5 µg/l | <5 µg/l | 160-600 ng/g |
| As | <5 µg/l | <5 µg/l | |
| Tl | <20 µg/l | | |
| Cd | - | <2 µg/l | |
| Pb | - | 50-150 µg/l | |
| Pt | <20 µg/l | | |
| Li* | 0.5-1.0 mmol/l | | |

*Therapeutic region

5. If the element analysis is done for purpose of diagnosis of disease, it has to be considered that in case of disease the above mentioned parameters have to be used carefully. In case of disease the daily requirements of elements may be altered and the recommended daily allowances may not meet the needs of the patient anymore. Fig. 1 shows the daily urinary losses of zinc and the therapy concept of a patient with severe zinc deficiency. The zinc deficiency developed during long term parenteral nutrition with parenteral food containing an amount of zinc according to the RDA. Obviously, the daily recommended zinc intake did not meet the zinc requirement of the patient mostly because of the high urinary losses. The patient developed typical clinical symptoms for severe zinc deficiency as summarized under topic 7.

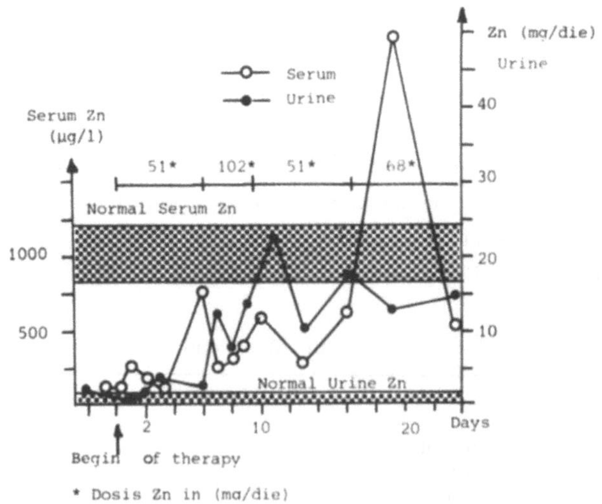


Fig. 1: Graphic presentation of the therapy concept of a patient with severe zinc deficiency due to longterm parenteral nutrition.

What can be learned from fig. 1 (12):

a) In case of disease, the RDA may be too low especially if losses via urine or feces occur.
b) The therapeutic dosis for the deficiency should be established by determination of the losses in case of parenteral nutrition via urine and in other nutritional supply and, if possible, also via feces.
c) The control of the therapy should be done by determination of the serum levels.

6. Besides the element concentrations in body fluids it is useful to obtain biochemical parameters associated with element deficiency or element intoxication. In case of zinc e.g., there are significant correlations between serum zinc concentration and vitamin A concentration (fig. 2). In zinc deficient patients, the retinol binding protein is increasing with the serum zinc concentration under zinc therapy (fig. 3). This phenomenon can be used as a control for the success of the therapy. One can conclude from this:

a) It is useful to control the therapy success in element deficiency by biochemical parameters. In copper deficiency it is recommended to measure the copper transporting protein in serum coeruloplasmin. If iron deficiency is treated by iron supplementation, it is necessary to determine ferritin concentration in serum to control the therapy.

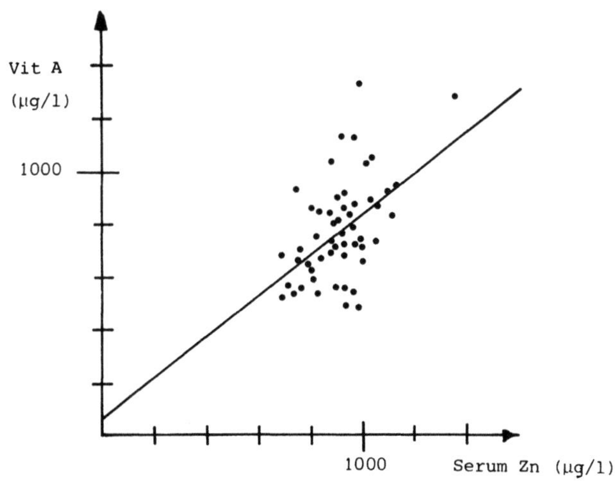


Fig. 2: Correlation of serum zinc with serum vitamin A concentration; regression equation $y = 0.78\ x + 63$, $r = 0.5$, $p < 0.001$.

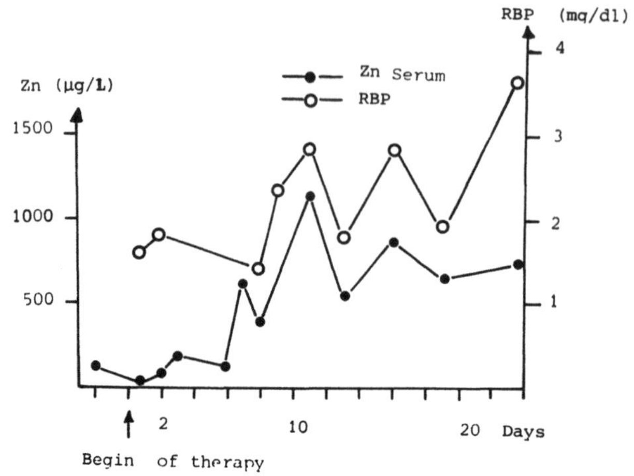


Fig. 3: Correlation of serum zinc with retinol binding protein (RBP) concentration

b) In cases of zinc deficiency, the zinc concentration in serum should be always looked at in combination with vitamin A and the retinol binding protein concentration, and vice versa. It is thus advisable to determine also the zinc status in case of low vitamin A and low retinol binding protein concentration in serum. In case of low zinc and low vitamin A concentration, the author recommends a zinc supplement in order to stimulate the retinol binding protein secretion and the vitamin A secretion by the liver. In some cases, the intramuscular supplementation of vitamin A is not necessary and the danger of hypervitaminoses A can be prevented (12,13).

7. As already mentioned, the elements are taken up by food. If the food supply does not provide enough of the essential elements, deficiency symptoms develope. For the diagnosis of deficiency by element analysis, the deficiency symptoms have to be known. The deficiency symptoms for the elements Na, Cl, K, Ca, P, Mg, Fe, Cu, Zn, J, F, Se, Cr, Co and Mn are:

Sodium: Na deficiency is associated with a decrease of blood pressure, tendency to collaps, oliguria, apathy, fever, dizzyness, malaise.

Chlorine: cf sodium

Potassium: Neuromuscular symptoms with adynamia and, in severe cases, with paralysis of the skeletal muscle. Also observed are apathy, somnolence and coma.

Calcium: For the resorption of Ca, both vitamin D and the conversion of vitamin D in the kidney into 1,25-dihydroxycholecalciferol are essential. The resorption of Ca is also influenced by the amount of phosphate in the food. Clinical symptoms are osteoporosis, paralysis and, in very severe cases, also hemorrhagies. If the calcium deficiency is due to vitamin D deficiency, ricketts is the predominant clinical symptom.

Phosphorus: No specific symptoms, e.g. adynamia.

Magnesium: Tremor, tetany, tachycardia, hypertension and adynamia.

Iron: Nonspecific clinical manifestations of iron deficiency anemia are fatigue, pallor, tachycardia, tinnitus, headaches. In addition to these symptoms, iron deficiency causes characteristic alterations in the structure and cytology of various organs:
Skin and mucous membrane: glossitis and erythema at the tip of the tongue, angular stomatitis (cheilosis). The red glazed tongue in iron deficiency is usually painless, but may exhibit painful vesicles or ulcers. Abnormal cornification and frequency of subepithelial inflammation of the buccal mucose. Ocurrence of koilonychia is rare in iron deficiency.
Alimentary tract: Atrophic changes occur in the esophagus, in the gastric mucosa also a high frequency of atrophic gastric mucosal changes are observed. In children with iron deficiency anemia, the small intestinal mucosa changes are consisting of widening, blunting and fusion of villi and submucosal inflammatory reaction, changes similar to those seen in sprue.
Blood: Only in very advanced iron deficiency states morphological changes of the erythrocytes are observed associated with a hypochromic microcytic anemia. In addition to microcytosis, the erythrocytes typically exhibit elliptocytosis.
Bone: Growth retardation and skeletal abnormalities may occur in iron deficiency.

Copper: Clinical symptoms in the human infant are anemia, hypoproteinemia and low serum iron levels. In infants, the copper deficiency may be due to the low copper content of the milk. In malnourished infants being rehabilitated on high calorie diet, which is low for Cu, anemia, neutropenia, chronic or recurrent diarrhea and "scurvy-like" bone changes are observed which are responsive to copper therapy.
Inborn diseases: The low copper and ceruloplasmin concentration in Wilson and Menkes disease is not due to copper deficiency in the diet but is an inborn error which shows abnormal copper distribution in the body.

Zinc: Zinc deficiency may be associated with anemia, loss of appetite, disturbances of smelling and taste, bullous pustular dermatitis of the extremities and the oral, anal and genital region. Woundhealing problems are typical for zinc deficiency as well as disturbances of the coagulation. Growth retardation with deleted puberty and hypogonadism. Zinc deficiency may be associated with hypovitaminosis A and night blindness. In severe cases both hair loss and low serum protein concentrations are typical. Because of the involvement of zinc in the protein biosynthesis a weakness of the immunosystem is observed. Neuropsychiatric symptoms are irritability, tremor and ataxia.
Inborn error: Acrodermatitis enteropathica is not a disease due to low zinc in the diet, but an inborn disease due to decreased absorption of zinc from the diet.

Iodine: In dietary deficiency of iodine enlargement of the thyroid gland or goiter is common. The enlargement of the thyroid gland or goiter however may not only be due to iodine deficiency but also to the presence of goitrogenes or to a constitutional disability. Endemic severe goiter is frequently associated with a condition known as endemic cretinism, which is characterized by a wide range of clinical abnormalities including mental retardation, deafness and deaf-mutism, retarded growth and neurological abnormalities, as well as hypothyroidism. There is evidence that iodine deficiency of the mother during the first third of pregnancy is the main factor for endemic cretinism and that iodine repletion prior to conception can prevent the condition. In man, colloid goiter often develops at puberty and hyperthyroidism is sometimes precipitated at menopause. Goitrous cretins are usually sterile and invariably fail to develop normal sexual vigor, with a delayed maturation of the genitalia.

Fluorine: Deficiency symptoms are not known but the addition of fluorine to drinking water is considered as a caries prophylaxe. Beneficial effects have been obtained in the treatment of patients with osteoporosis and other demineralizing diseases with NaF.

Selenium: Selenium deficiency in food is associated with an endemic cardiomyopathy in China (Keshan disease). Children and women of child bearing age are primarily affected. The cardiomyopathy is not associated with any specific symptoms. In the acute state, sudden insufficiency of the heart function is typical whereas in chronic subacute phases the disease is characterized by heart enlargement. Another disease associated with dietary selenium deficiency is the Kashin-Beck disease. Kashin-Beck disease is an endemic osteoarthopathy which also occurs in China (northern part), Siberia and North Korea. It is characterized as a chronic, disabling, degenerative, generalized osteoarthrosis that involves both the peripheal joints and the spine. There is disturbed enchondral ossification and deformity of the affected joints but no systemic or visceral manifestations.

Chromium: As mentioned above, chromium is involved in the glucose metabolism and chromium deficiency may be associated with elevated blood glucose concentrations.

Cobalt: Clinical symptoms due to cobalt deficiency are not described. It has to be mentioned that cobalt is an integral part of vitamin $B_{12}$ and that vitamin $B_{12}$ deficiency of course causes clinical symptoms which are not discussed here.

Manganese: Reports about clinical symptoms for Mn deficiency are not very profound. There seems to be an association between Mn deficiency and coagulation. The efficiency of vitamin K on the improvement on coagulation seems to be controlled by manganese. In Mn deficiency, a marked reduction of cholesterol, triglycerides and phospholipids are observed.

8. Due to the contamination of the environment by toxic elements, the widespread use of chemicals containing toxic elements in manufacturing the involuntary intake of toxic elements by accident, the intake of toxic elements by suicide, and the use of inorganic compounds as drugs should be taken into account the clinical symptoms of element intoxications (mostly metal and metalloid intoxications) for diagnostic purposes by element analysis.

Arsenic: The acute clinical effects in arsenic poisoning vary whether As is taken in as inorganic or organic compound. Symptoms may appear very short after ingestion or may be delayed (10-12 hours). Initially sharp crampy abdominal pains, vomiting and diarrhea is reported. The severe gastroenteritis is associated with cardiovascular collapse. Chronic arsenic poisoning is associated in the first stage with digestive symptoms of vomiting and diarrhea. In a later stage of intoxication conjuctivitis, rhinitis, laryngitis and bronchitis are common. Together with theses symptoms various skin eruptions appear. They are followed by neurological disorders with symptoms including paresthesia, hyperesthesia and neuralgias. Muscle tension was consistently noted. Mental confusion may be observed. In severe cases, changes of the blood vessels of the skin are present and painful, swollen, red extremities are found. The skin is characterized by pigmentation mostly around scars, the neck, in the armpits and around the nipples. The fingernails show Mess lines in severe cases. It has also to be mentioned that arsenic may be teratogenic, mutagenic and carcinogenic.
The diagnosis of arsenic poisoning may be done by the determination of arsenic in urine, whole blood, hair, stomach content and finger nails. Arsine intoxication can be

considered a special case of arsenic poisoning because of the high toxicity of $AsH_3$. The initial symptoms, usually of a vague, constitutional nature with headache, malaise, weakness, dizzyness and dyspnea are accompanied by gastrointestinal disturbances, with abdominal pain, nausea and vomiting. The occurrence of dark red urine is frequently noticed, usually four to six hours after inhalation of the gas. After 24 to 28 hours, jaundice is present. Physical examinations may reveal fever, tachycardia and tachypnea. A peculiar bronze tint of the skin has been described. The most important laboratory finding is a hemolytic anemia with low whole blood hemoglobin and increased free hemoglobin in the plasma due to rapid hemolysis of the erythrocytes. The Coombs test however is negative. In urine, the hemoglobin content is increased. The diagnosis has to consider that massive hemolysis may also occur in other diseases.

Cadmium: The clinical symptoms of acute poisoning are not that good described than the chronic effect of cadmium intoxication. The acute effects consist of prompt nausea and vomiting. If the cadmium is inhaled, acute pneumonitis may be observed. The chronic effect of a cadmium intoxication is dependent on the duration of the intoxication. First, a decreased calcium resorption is described followed by hypertension and tubular kidney damages. An increased urinary calcium excretion is typical. In severe cases, kidney stones anemia and osteomalacia/osteoporosis may occur.
Laboratory data which give information about intoxication are the cadmium concentration of whole blood and urine. Especially important are laboratory data describing the kidney function (tubular damages can be detected by the determination of $\beta$-2-microglobulin) and the parameters describing calcium metabolism (24 hours calcium excretion in urine and phosphate clearance).

Mercury: In mercury intoxication, it has to be differentiated between the intoxications with inorganic and organic mercury compounds. The intoxication with Hg vapor reveal symptoms like metal vapor feaver. The inhalation acute of a high concentration of elementary mercury vapor may give rise to symptoms of pulmonary irritation together with involvement of the nervous system. Cough, bronchial irritation, chest pain and rapid respiration with excitement and tremor have been described. The longterm inhalation of mercury vapor give rise to classical mercurialism. The characteristic features of mercurialism are gingivitis and salivation, tremor and erethism.
Ingestion of soluble mercuric compounds initially causes upper abdominal pain as well as vomiting. The initial symptoms are followed by severe diarrhea with the passage of blood and necrotic epithelium, which may be accompanied by circulatory collapse and death. In severe cases, kidney damages and neurological disorders are very common in mercurialism.
The primary signs of intoxication with methyl mercury compounds result from damage of the nervous system and are characterized by loss of sensation at the tips of the hands and feet and in areas around the mouth (paresthesia), loss of coordination in gait, dysarthia, concentric constriction of the visual field and loss of hearing, severe poisoning may cause blindness, coma and death. Prenatal exposure to methyl mercury results in mental retardation with cerebral palsy.
Acrodynia or pink disease is a special case of mercury poisoning. Children are especially affectet. The acrodynia is a toxic or toxic allergic effect of mercury on the sympathicus in the mesencephalon. The children become irritable, sleepless, perspired profusely and developed photophobia. A generalized rash follows, accompanied by coldness, pain and swelling of the extremities. Desquamation and ulceration is observed.
The diagnosis in all cases of mercury intoxication is done by the determination of mercury in urine, serum and whole blood. If the intoxication happened with organic mercury compunds, the determination of mercury in the erythrocytes should be done. Additionally, to detect kidney damages, the parameters describing the kidney function have to be determined.

Lead: Inorganic lead poisoning in adults is mostly associated with symptoms referring to the abdomen and the triad of colic, nausea and vomiting. Neurological signs are peripheral paralysis with decreased nerve conduction. Encephalopathy in adults is unusual, except at continued heavy exposure. The major features of this condition are irritability, tremor and hallucinations, headaches, restlessness, loss of memory and in-

ability to concentrate. Inorganic lead intoxication causes renal damages with chronic nephropathy, proximal tubular dysfunction and gout.
In children, lead poisoning is more often associated with encephalopathy. Inorganic lead poisoning may cause psychotic changes with the symptoms insomnia, delirium, hallucinations and tremor.
Laboratory data for the detection of lead intoxication are especially the whole blood lead concentration and the urinary lead excretion. In cases of organic lead poisoning, the total blood lead concentration may be normal, but the lipid containing fraction is usually elevated. In cases of organic lead intoxication, no elevation of the urinary excretion of ALA of coproporphyrin is observed as in inorganic lead poisoning. In organic lead poisoning,red cell protoporphyrin is only moderately raised compared to inorganic lead intoxication. Lead poisoning usually is associated with severe anemia. Therefore, the determination of the hemoglobin concentration, the number of erythrocytes and the hematocrit is necessary.

Thallium: Soon after acute thallium poisoning, vomiting is typical followed by a rather 2-4 symptomless days with obstipation. After 2-4 days, painfull obstipation is found. Parallel neurological symptomatology develops, consisting of dysarthria, neuropathy, granular nerve palsy and coma. The lower extremities are usually painful, which suggests that smaller peripheral nerves known to be favorable for conduction of sensory rather than motor impulses are preferentially affected by the toxin. Motor neuropathy is followed by weakness. The gastrointestinal symptoms during this time are associated with bleeding and ileus. Alopecia described as typical in thallium intoxication occurs rather late during intoxication at about 10-15 days after intoxication, which is too late to be used for diagnostic purposes.
The laboratory data necessary for the diagnose of thallium intoxication are to be determined as early as possible by analysis of thallium content in serum and urine.

9. Besides the diagnosis of deficiency and intoxication, the element analysis of body fluids may give a broad spectrum of information ranging from vitamin deficiency, evaluation of the hormonal status, being involved in tumor diagnosis, giving information about inflammatory diseases and organ function diagnosis. In summary, the indications of pathophysiological conditions the element analysis may give information about the importance of the element analysis for the diagnosis of diseases is again emphasized. Indications other than deficiency and intoxication for the determination of elements in human body fluids are as follows:

Sodium: Disturbances of the electrolyte and water balance, diarrhea, polyuric and polydiptic syndrome, disturbances of the acid-base-metabolism, hypertension, kidney insufficiency, hypothyreose, intravenously nutrition, infusion therapy.

Chlorine: cf. sodium.

Potassium: Disturbances of the acid-base-metabolism, hypertension, digitalis therapy, diuretica and laxative therapy, acute and chronic kidney insufficiency, heartarythmia, infusion therapy.

Calcium: Bone pain, kidney stones, spontaneous fracture of bones, polydipsie and polyurie, acute and chronic kidney insufficiency, malignant tumors, tetany, Morbus Boeck, vitamin D deficiency, vitamin D therapy control, delayed growth, gammopathy, anorexia, weight loss, muscle weakness, hyper- and hypoparathyreoidism, lethargy, depression, diarrhea, obstipation, abnormal fingernails and teeth, cortison therapy.

Magnesium: Neuromuscular irritability, tremor, tetany, tachycardia, arrythmia, changes in the electrocardiogram, alcoholism, diuretica therapy, in parenteral nutrition.

Phosphorus: In bone diseases, chronic kidney insufficiency, kidney stones, parenteral nutrition, hyper- and hypoparathyreoidism, disorders of the thyroid gland, dysparathyreoidism.

Nitrogen: In parenteral nutrition determination of nitrogen in the urine gives information on the nitrogen balance.

Iron: In inflammatory diseases, in malignant tumors together with ferritin, iron overloading.

Copper: Morbus Wilson, Menkes syndrome, liver diseases, icterus, inflammatory diseases, malignant tumors.

Aluminium: Kidney insufficiency with dialysis, aluminium therapy.

Lithium: Lithium therapy.

10. If the diagnosis of diseases is done by element analysis, it has to be considered that changes of the element concentrations in body fluids may not only result from deficiency and intoxication, but may also be of iatrogen origin. Under the therapy of digitalis, cortison, diuretica, laxatives, in parenteral nutrition and infusion therapy etc., the element concentrations in body fluids may be altered.

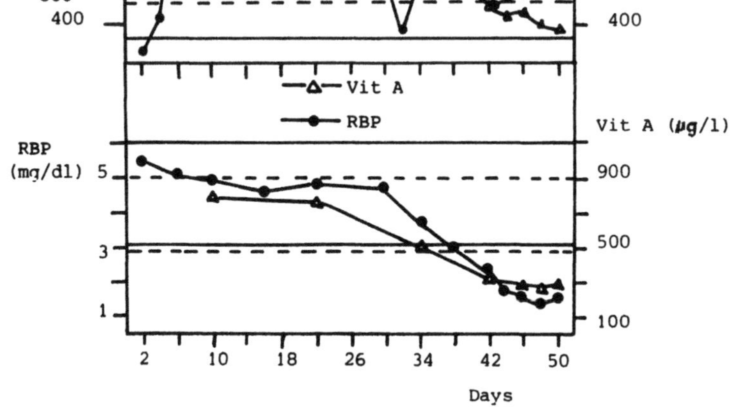


Fig. 4: Decrease of the serum zinc, the serum vitamin A, the RBP concomitant with increased urinary losses of zinc during DMPS therapy.

Iatrogen changes of element concentrations in body fluids may happen also if drugs are applied which contain inorganic compounds as pharmaceutics. The elements Ca, Mg, Na, K, P, Fe, Cu, Zn, Cr, Li, Bi, Co, Se, Mn, J, F, Hg, Pb, Al, As etc. are used as drugs. To avoid an accumulation of the element in the body, the element control in body fluids should be done.

Another iatrogen change in the element concentration of the body fluids is the increased excretion of essential trace elements by the use of chelators as drugs e.g. penicillamin, captagon, DMPS etc. In Fig. 4, the development of zinc deficiency during DMPS therapy due to mercury poisoning is documented as an example of iatrogen changes in the element composition of the human body.

## Conclusion

The element analysis used for the diagnosis of diseases includes the major and the essential trace elements as well as the toxic elements. However, it has to be considered that the toxicity of an element is always a question of the amount present in the body. That means that every element can be toxic in amounts the organism is not adapted to.

The element analysis can give information upon:

a) deficiency of an element in the food supply
b) intoxication of an element by food, air, skin contact etc.
c) the success of the therapy of deficiency and intoxication
d) pathophysiological conditions
e) the therapy of patients treated with drugs containing inorganic substances.

The information obtained by element analysis however can be interpretated adequately only if a basic knowledge upon the biochemical function of the elements in the human organism is present as well as upon daily requirements of the elements and upon daily losses and the toxicity of the elements.

## References

1) ICRP Publication 23 (1974) Report of the Task Group on Reference Man. Pergamon Press, Oxford.

2) Oster O, Prellwitz W (1986) Die Anwendung der Atomabsorptionsspektrometrie im klinisch chemischen Laboratorium. In: Welz B (ed) Fortschritte in der atomspektrometrischen Spurenanalytik. VCH Verlagsgesellschaft, Weinheim. pp 186-195.

3) Oster O, Prellwitz W (1984) Die Bestimmung toxischer Spurenelemente für die Akut-Diagnostik. In: Welz B (ed) Fortschritte in der atomspektrometrischen Spurenanalytik. VCH Verlagsgesellschaft, Weinheim. pp 275-284.

4) Oster O, Prellwitz W (1985) Die Pathobiochemie, Diagnose und Therapie der Metall- und Metalloidintoxikation - Teil 1. Intensivmedizin 22: 51-60.

5) Oster O, Prellwitz W (1985) Die Pathobiochemie, Diagnose und Therapie der Metall und Metalloidintoxikation - Teil 2. Die Quecksilberintoxikation. Intensivmedizin 22: 130-139.

6) Oster O, Prellwitz W (1985) Die Pathobiochemie, Diagnose und Therapie der Metall- und Metalloidintoxikation - Teil 3. I. Die Arsenvergiftung/II. Die Thalliumvergiftung. Intensivmedizin 22: 276-285.

7) Recommended Dietary Allowances, Ninth Revised Edition (1980), National Academy of Sciences, Washington D.C.

8) Deutsche Gesellschaft für Ernährung (1986) Empfehlungen für die Nährstoffzufuhr. 4. erweiterte Überarbeitung, Umschau Verlag, Frankfurt.

9) Deelstra H, van Schoor O, Robberecht H, Clara R, Eylenbosch W (1987) Daily chromium intake by infants in Belgium. Europ J Pediatr (in press)

10) Kumpulainen J, Vuore MD (1980) Longitudinal study of chromium in human milk. Amer J Clin Nutr 33: 2299-2302.

11) Casey CE, Hambidge KM (1984) Chromium in human milk from American mothers. Brit J Nutr 52: 73-77.

12) Oster O, Prellwitz W (1984) Die Notwendigkeit der Bestimmung von Spurenelementen im klinisch chemischen Laboratorium. Ärztliches Laboratorium 30: 119 127.

**PROPOSALS AND RECOMMENDATIONS OF THE CONFERENCE CONCERNING FUTURE WORK ON TRACE ELEMENTS**

Gisela Grupe (editorial revision)

The participants of the conference agreed that trace element analysis of ancient human tissues means a great research potential in prehistoric anthropology and human environmental history. Although a lot of problems still do exist, possibilities for their solutions have been figured out during the conference. Thus, it is hoped that common research strategies in Europe will work to enhance and facilitate progress in this field of research.

In a concluding session, the following statements and recommendations were worked out by the participants:

1) It was generally agreed that trace element analysis is a very complex matter. But it has been also generally agreed that it serves as a tool for reconstructing nutritional habits as well as other environmental factors, e.g. occupational diseases, polluting agents, soil- and groundwater features.

2) With respect both to the reconstruction of ancient food chains and to bioavailability of trace elements soil ingestion, which is a bypass of the natural food chain, has to be taken into account. It has also been found that trace element contents differ in various plant tissues, thus complicating the evaluation of discrimination factors. Furthermore, individual preference as well as social factors are likely to influence the "observed ratios", which are generally used to calculate the discrimination factors. As a consequence, the reconstruction of dietary habits should not be calculated from a single, distinct observed ratio, but a wider range of this ratio should be used instead.

3) For preindustrial times, trace element intake is mainly or almost exclusively based on the consumption of food and drinking water. However, for certain conditions as in mining areas, exposure to smoke from firing places and for special occupations a trace element intake via the lung or skin may be considered. When the trace element contents in ancient human tissues are measured, heavy metals should be included to provide a baseline for modern ecomedicine.

4) Differential suitability of various preserved human tissues was discussed. Most of the data have been obtained from bone. Trace element analysis of teeth is promising, whilst analysis of prehistoric hair is not yet fully understood due to lack of material. It is recommended to apply trace element analysis also to mummified soft tissue remains. However, hair and soft tissue studies are unlikely to be of use for cross sectional studies in most cases, since these tissues are rarely preserved. An advantage of analysing all kinds of preserved tissues whenever possible will lie in the covering of different individual time spans of trace element intake.
Since modern clinical diagnosis of trace element status is based on body fluids, knowledge on the relations of trace element contents between body fluids and hard tissues is supposed to be valuable. It might be easier to obtain these data for teeth than bone.

5) Many published data cannot be compared directly due to a lack of standardised methods concerning specimen sampling and preparation. It was generally agreed that compact tissue is most recommended as far as bone is concerned. Compact tissues are preserved in most cases and easily available. They are also less susceptible to diagenetic alterations and variation in trace element content is minimised in shafts of any long limb bone. Certainly realization of the suggestion depends on the availability of material. But when different kinds of bone are analysed, intraindividual variation in trace element contents have to be taken into account before any conclusions on average values are drawn.
It was emphasised that all published data should include the result of the standard used for calibration for purposes of quality control.

6) No standardised procedures can overcome the problem of the geographical variability of trace element distributions. Parallel analysis of faunal remains from the same site and of animals which fed on known diets can provide a baseline for comparison among samples from different regions.
Parallel analysis of soil samples from the same site the human remains come from are highly recommended, both for the assessment of diagenetic changes and for the facilitation of comparison among samples of different geographical origin.

7) Experimental works on modern animals could make clear some problems such as the intraindividual variability and the relative contribution of the trace element contents in bone of few investigated foodstuffs (e.g. honey, berries, molluscs etc.).

8) Future research should thus especially focus on the investigation of modern animal populations. It should also concentrate on experimental work on contamination and loss of trace elements in biological materials under different conditions.

## SUBJECT INDEX

MIX
Papier aus verantwortungsvollen Quellen
Paper from responsible sources
FSC® C105338

If you have any concerns about our products,
you can contact us on
**ProductSafety@springernature.com**

In case Publisher is established outside the EU,
the EU authorized representative is:
**Springer Nature Customer Service Center GmbH**
**Europaplatz 3, 69115 Heidelberg, Germany**

Printed by Libri Plureos GmbH
in Hamburg, Germany